T0192284

# Lecture Notes in Computer Science 14525

## Founding Editors

Gerhard Goos

Juris Hartmanis

## Editorial Board Members

The series Lecture Notes in Computer Science (LNCS), including its subseries Lecture Notes in Artificial Intelligence (LNAI) and Lecture Notes in Bioinformatics (LNBI), has established itself as a medium for the publication of new developments in computer science and information technology research, teaching, and education.

LNCS enjoys close cooperation with the computer science R & D community, the series counts many renowned academics among its volume editors and paper authors, and collaborates with prestigious societies. Its mission is to serve this international community by providing an invaluable service, mainly focused on the publication of conference and workshop proceedings and postproceedings. LNCS commenced publication in 1973.

Éric Renault · Selma Boumerdassi ·
Paul Mühlethaler
Editors

# Machine Learning for Networking

6th International Conference, MLN 2023
Paris, France, November 28–30, 2023
Revised Selected Papers

 Springer

*Editors*
Éric Renault
ESIEE Paris and Gustave Eiffel University
Noisy-le-Grand, France

Selma Boumerdassi
Conservatoire National des Arts et Métiers
Paris, France

Paul Mühlethaler
Inria
Paris, France

ISSN 0302-9743          ISSN 1611-3349 (electronic)
Lecture Notes in Computer Science
ISBN 978-3-031-59932-3          ISBN 978-3-031-59933-0 (eBook)
https://doi.org/10.1007/978-3-031-59933-0

This Springer imprint is published by the registered company Springer Nature Switzerland AG
The registered company address is: Gewerbestrasse 11, 6330 Cham, Switzerland

If disposing of this product, please recycle the paper.

# Preface

The rapid development of new network infrastructures and services has led to the generation of huge amounts of data, and machine learning now appears to be the best solution to process these data and make the right decisions for network management. The International Conference on Machine Learning for Networking (MLN) aims to provide a top forum for researchers and practitioners to present and discuss new trends in machine learning, deep learning, pattern recognition and optimization for network architectures and services. MLN 2023 was jointly organized by the EVA Project of Inria Paris, the Laboratoire d'Informatique Gaspard-Monge (LIGM) and ESIEE Paris of Université Gustave Eiffel, and CNAM Paris, and was hosted by Inria Paris, France.

The call for papers resulted in a total of 34 submissions from all around the world: Algeria, Canada, Egypt, France, Morocco, Saudi Arabia, South Africa, Tunisia and the USA. All submissions were assigned to at least three members of the program committee for double-blind review. The program committee decided to accept 16 papers. Two invited papers were also presented.

The paper *Data Summarization for Federated Learning* by Julianna Devillers, Olivier Brun and Balakrishna Prabhu from ISAE and LAAS-CNRS, France, was awarded the prize for the best paper.

Four keynotes completed the program: *Building Digital Twins for Smart-cities and Intelligent Infrastructures* by Ahcene Bounceur from KFUPM University, Dhahran, Saudi Arabia, *Large Language Models for wireless: The Next Revolution* by Merouane Debbah from Khalifa University, Abu Dhabi, UAE, *IoT Smart Systems for Livable Smart Cities and Communities* by Mohsen Guizani from MBZUAI, UAE, and *The Art of the Possible with Machine Learning in IoMTs* by Osman Salem from Université Paris Cité, France.

We would like to thank all who contributed to the success of this conference, in particular the members of the Program Committee and the reviewers for carefully reviewing the contributions and selecting a high-quality program. Our special thanks go to the members of the Organizing Committee for their great help.

We hope that all participants enjoyed this successful conference.

January 2024

Paul Mühlethaler
Éric Renault

# Organization

## General Chairs

Paul Mühlethaler                 Inria, France
Éric Renault                   ESIEE Paris – Université Gustave Eiffel, France

## Steering Committee

Selma Boumerdassi           CNAM, France
Éric Renault                   ESIEE Paris – Université Gustave Eiffel, France

## Organization Committee

Lamia Essalhi                ADDA, France

## Technical Program Committee

Aissa Belmeguenai          Université de Skikda, Algeria
Fred Aklamanu              Nokia Bell Labs, France
Selma Boumerdassi          CNAM, France
Aravinthan Gopalasingham    Nokia Bell Labs, France
Viet Hai Ha                 Hue University, Vietnam
Cherkaoui Leghris            Hassan II University, Morocco
Ruben Milocco              Universidad Nacional del Comahue, Argentina
Paul Mühlethaler            Inria, France
Éric Renault                   ESIEE Paris – Université Gustave Eiffel, France
Van Khang Nguyen          Hue University, Vietnam
Mounir Tahar Abbes         University of Chlef, Algeria
Van Long Tran              Enlab Software, Vietnam

# Sponsoring Institutions

CNAM, Paris, France
ESIEE, Paris, France
Inria, Paris, France
Université Gustave Eiffel, France

# Contents

# Machine Learning for IoT Devices Security Reinforcement

Philippe Ea, Jiahui Xiang, Osman Salem$^{(\boxtimes)}$, and Ahmed Mehaoua

Centre Borelli, UMR 9010, Université Paris Cité, Paris, France
{Philippe.ea,jiahui.xiang,osman.salem,ahmed.mehoua}@u-paris.fr

**Abstract.** As more lightweight objects connect to the Internet, the Internet of Things (IoT) is changing our linked environment. Thus, IoT intrusion detection research must be high-quality to provide solutions. Network intrusion datasets are essential for training and testing attack detection algorithms. This study describes, statistically analyzes, and machine learning evaluates the innovative ToN IoT dataset. Heterogeneity in IoT datasets is important and can affect detection performance. We demonstrate through a cross-validated experiment with five classifiers that industry-useful IoT network intrusion datasets require several data collection methods and a wide variety of monitored variables. Among our five models, RF emerged as the best one achieving an impressive accuracy of 0.991 with an Area Under the Curve of 0.9996. Moreover, its training time took only 0.985 s. Our study reviews an important amount of datasets focused on IoT security. We will see that standardizing feature descriptions and cyberattack classifications is necessary for the operational use of IoT datasets.

**Keywords:** Intrusion detection · Internet of Things · Machine Learning · Deep Learning · Network security

## 1 Introduction

Internet of Things (IoT) devices have enabled the rapid proliferation of Internet-connected electronics. Since smart devices become less costly and simpler to produce, the IoT business grows daily [11]. IoT applications have developed quickly [10,18]. They play a major role in automotive innovations for example. However, this intensive usage leads to security issues: Palo Alto Networks' global threat intelligence team's 2020 IoT Threat Report, in [1], argues that 98% of IoT data is unencrypted, disclosing personal data and that IoT devices often run obsolete software, making them vulnerable. Device vulnerabilities, malware, and user habits such as password reuse are the biggest IoT concerns, according to this report. Another work in [2] categorizes IoT security threats and studies advanced IoT intrusion detection methods. New signature-based, anomaly-based, and specification-based models are used. Validating novel detection algorithms is difficult without high-quality labeled datasets as detailed

E. Renault et al. (Eds.): MLN 2023, LNCS 14525, pp. 1–13, 2024.
https://doi.org/10.1007/978-3-031-59933-0_1

in [17]. The key obstacles in creating such datasets are the fact that conventional IoT devices generate substantially less traffic than workstations and servers in conventional networks, and the ongoing introduction of new "objects" on the Internet, which creates fresh traffic and assaults. Heterogeneous datasets with diverse IoT devices, traffic, and assaults will be necessary. Anomaly detection techniques can identify malicious IoT device traffic in heterogeneous datasets.

Most IoT devices have low network traffic. For instance, sensor data, such as temperature variations, is sent from devices. Log and raw data are included. There are three data types:

1. Sensor data: IoT devices have sensors, functional software, and network connectivity. Smartwatches, door locks, temperature regulators, and voice controllers all provide sensory data via a network to regulate applications. Sensor data may show smart device (mis)use, hence it must be tracked.
2. Raw data: Every data packet sent includes sensor data. through a network by an IoT device includes connection and network protocol information. This raw data regarding these IoT devices' internal and exterior behavior complements sensor data.
3. Log data: Network analyzer PCAP data files are big and difficult to interpret. Today, security analysts employ software like Zeek (formerly known as Bro) to unobtrusively examine network traffic, interpret these observations, and build compact transaction logs with the results.

The network used to produce the dataset should have many distinct devices and network technologies, and the intrusion detection dataset should reflect this heterogeneity in its traffic data and attack types. Heterogeneity is important because the rapid growth of IoT in recent years is partly due to the creation of new technologies that simplify the installation and use of networked devices, resulting in an extensive selection of standards and solutions at each network layer and application stack.

IoT network intrusion datasets usually employ PCAP files for network data. The vast majority of network intrusion detection algorithms for IoT systems are tested by recreating intrusions on relatively old datasets like KDD-99 and NSL-KDD [5], UNSW-NB15 [14], or CICIDS [16]. Modern IoT networks use many protocols, standards, and technologies, which these data fail to cover. IoT Network Intrusion Dataset [8], Aposemat IoT-23 [7], and N-BaIoT [12] may address this weakness and provide a benchmark for efficient detection technique comparison. However, as shown later, they still analyze a limited collection of IoT devices and attacks. Therefore, we will focus on the novel ToN IoT dataset from [3, 6, 13]. Firstly, the dataset will be described and analyzed. It consists of devices and various attack types from PCAP files, Bro logs, sensor data, and Operating System (OS) logs. We then preprocess the data and use several Machine Learning (ML) models to find the most optimal performance. This study demonstrates how dataset heterogeneity enhances IoT intrusion detection based on ML.

The rest of this paper is organized as follows. Section 2 covers public IoT datasets. Section 3 describes the ToN IoT dataset we used, its collection, and its characteristics. We also explain the classifiers used in our study. Section 4 covers the results obtained after our classifier tests. Finally, Sect. 5 concludes our work.

## 2 Related Work

Since digital networks formed, datasets containing both anomalous and benign traffic were required for intrusion detection automation. These datasets allow security researchers and analysts to compare benign and malicious traffic patterns using ML. Publicly available datasets abound. Most of these consider generic network traffic (e.g. NSL-KDD [5] and CTU-13 [19]), while few are built for anomaly detection in IoT networks.

Recently available datasets for intrusion detection in IoT networks have different purposes and data kinds. We will discuss these datasets' heterogeneity here. Recent and heterogeneous datasets are considered: DS2OS traffic traces [15], Bot-IoT [9], IoT Network Intrusion Dataset [8], N-BaIoT [12] and Aposemat IoT-23 [7].

The DS2OS traffic traces dataset was released on Kaggle in [15] and [4]. It is DS2OS-generated synthetic data from a virtual IoT environment. The IoT system architecture uses Message Queuing Telemetry Transport (MQTT) to communicate between microservices. The dataset includes 13 attributes from monitoring connections between 7 Virtual State Layer (VSL) service types that connect illumination controls, movement sensors, thermostats, solar batteries, washing machines, door locks, and smartphones. This dataset contains microservice source, destination, operation, and other information but no network flows or packet data. It only has features that detect irregularities in IoT traffic frequencies and microservice communication baseline models. Thus, this dataset can only be used for a narrow class of detection and will be disregarded in our study.

The Bot-IoT dataset [9] helps identify IoT botnets. Both regular and attacking Virtual Machines (VM) were simulated in an IoT network. Kali VMs conducted port scanning, Distributed Denial of Service (DDoS), and botnet attacks. Node-red simulated weather stations, smart fridges, motion-activated lighting, garage doors, and thermostats on Ubuntu VMs. Argus extracted 46 network characteristics. No sensors or logs were recorded. Bot-IoT is the only dataset employing simulated IoT devices. The dataset is less typical of real IoT traffic because no real hardware is tracked. This led some of the BoT-IoT dataset producers to establish the ToN dataset, as explained in this study. Given the shortcomings of DS2OS and BoT-IoT, the paper will only compare ToN IoT with the IoT Network Intrusion Dataset, N-BaIoT, and Aposemat IoT-23.

The 2019 IoT Network Intrusion Dataset, in [8], included two smart home devices, an SKT NUGU (NU 100) and EZVIZ Wi-Fi Camera (C2C Mini O Plus 1080P), a few portable computers, and a few cellphones. Nmap was used to simulate assaults on this wireless network. It was one of the first IoT infiltration datasets published. The dataset contains 42 raw network packet files from different times. The dataset is homogeneous since no other data was obtained.

Another study in [12] introduces a network-based botnet detection dataset: N-BaIoT. The IoT devices were Wi-Fi connected to numerous access points and tethered to a central switch connected to a router. BASHLITE and Mirai were botnet attacks. N-BaIoT characteristics are statistical quantities produced from sensor logs from each device in the experimental setup, unlike the IoT Network Intrusion Dataset.

Finally, Aposemat IoT-23 is another IoT traffic and network monitoring dataset detailed in [7]. It captures IoT traffic from genuine hardware devices including a Philips HUE smart LED bulb, Amazon Echo, and Somfy Smart Door Lock. The Stratosphere experiment collected 20 malware samples from IoT devices and 3 benign samples. Mirai and Torii botnets were used. Zeek IDS netflows and labels are offered alongside packet capture files.

## 3   Proposed Approach

In this section, we present the five classifiers used in our study: Logistic Regression (LR), Decision Tree (DT), Random Forest (RF), and Multi-Layer Perceptron (MLP).

LR is a statistical method used for binary classification. Any input value is translated into a value between 0 and 1 using the logistic function, also referred to as the sigmoid function. As a result, the logistic regression model can produce the likelihood that the input belongs to a particular class. This model can be mathematically defined as follows:

$$P(Y = 1|X) = \frac{1}{1 + e^{-(\beta_0 + \beta_1 X)}} \tag{1}$$

$P(Y = 1|X)$ is the probability of the input belonging to class 1, $\beta_0$ and $\beta_1$ are the model parameters, and $X$ is the input feature vector. LR provides a straightforward way to estimate the likelihood of an input belonging to a particular class.

The DT model is a commonly utilized technique in the domain of ML, renowned for its straightforwardness and comprehensibility. The model above is a method utilized for addressing classification and regression problems. It takes the form of a tree and functions by dividing the data into subsets based on the value of a selected feature, through a process of recursive partitioning. The construction of a decision tree involves the identification of the feature that provides the highest information gain for partitioning the data, followed by the creation of branches for each potential outcome. The iterative procedure is continued until a halting condition is met, which could be either the prescribed limit on the depth of the tree or the minimum threshold for the number of samples in each leaf node. The tree that ensues can be readily visualized and interpreted, rendering it a favored option for diverse applications, including but not limited to medical diagnosis, fraud detection, and customer segmentation.

RF is a classification algorithm that employs numerous trees for training and predicting samples. These trees form multiple forests and generate an ensemble of decision trees in a randomized manner. Upon acquisition of the forest, after its construction, the introduction of a novel sample prompts each decision tree within the forest to independently evaluate and determine the appropriate classification for said sample. Based on the classification frequency, it is anticipated that the sample belongs to the class that it was predominantly assigned to. Mathematically, RF can be represented as:

$$\text{RF}(X) = \frac{1}{N} \sum_{i=1}^{N} \text{DT}_i(X) \tag{2}$$

With $N$ the number of trees in the forest and $\text{DT}_i(X)$ the prediction of the $i$-th decision tree. Random Forest is known for its robustness and is effective in handling complex datasets.

MLP is an Artificial Neural Network (ANN) that is extensively employed in the domain of Deep Learning. The architecture of this system is comprised of a series of concealed layers of interconnected neurons that are situated between the input and output layers. Every neuron within the MLP is a mathematical function that accepts input signals, performs a transformation utilizing a predetermined set of weights and biases, and generates an output. The process of training an MLP involves the acquisition of knowledge about the weights and biases, which are adjusted to minimize the discrepancy between the predicted output and the actual output. MLPs possess the ability to acquire intricate nonlinear associations within the data and are widely employed in diverse domains, including but not limited to image and speech recognition, natural language processing, and financial prediction. The architecture of a simple MLP can be represented as:

$$\begin{aligned}
Z^{(1)} &= \sigma(W^{(1)} X + b^{(1)}) \\
Z^{(2)} &= \sigma(W^{(2)} Z^{(1)} + b^{(2)}) \\
&\;\;\vdots \\
Z^{(L)} &= \sigma(W^{(L)} Z^{(L-1)} + b^{(L)})
\end{aligned} \tag{3}$$

where $Z^{(L)}$ represents the output of layer $l$, $W^{(l)}$ and $b^{(l)}$ are the weights and biases for layer $l$, and $\sigma$ is the activation function.

## 4  Experimental Results

Firstly we analyze the ToN IoT dataset. It refers to a set of 24 datasets that represent a novel iteration of IoT and Industrial IoT (IIoT) datasets. They assess the effectiveness and accuracy of various cybersecurity applications. These applications include but are not limited to intrusion detection, threat intelligence, adversarial ML, and privacy-preserving models.

The dataset was given the name "ToN IoT" due to its composition of diverse data sources, which were obtained from Telemetry datasets of IoT and IIoT sensors, OS datasets of Windows 7 and 10, Ubuntu 14 and 18 TLS, and network traffic datasets. This network was designed to incorporate many VMs, physical systems, hacking platforms, cloud and fog platforms, and IoT sensors, to emulate the intricate and expansive nature of IIoT and Industry 4.0 networks.

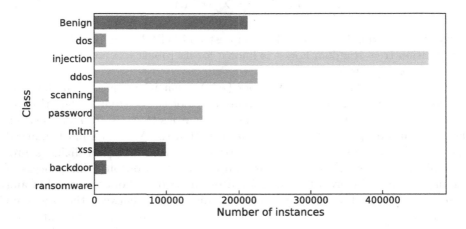

**Fig. 1.** Detailed Distribution of Classes

The architecture of the testbed comprises three distinct layers, namely the edge/IoT layer, the fog layer, and the cloud layer. Both the edge and fog layers provide on-premise services, such as cloud services, and enable the processing of extensive data sources, as well as the implementation of data analytics and intelligence close to end users. The primary distinction between the edge and fog layers pertains to the placement of intelligence, analytics, and processing power within the network architecture. Specifically, the edge layer involves the integration of these capabilities into devices, such as embedded automation controllers and lightweight IoT devices, whereas the fog layer involves the placement of these capabilities elsewhere within the Local Area Networks (LANs). A comprehensive exposition of this experimental platform can be found in the research conducted in [13].

The dataset covers several types of attacks as we can see in Fig. 1. The most prominent ones are injection, DDoS, cross-site scripting (XSS), and password-related attacks. In Fig. 2, we can observe that the attacks are overrepresented in the dataset. Therefore, we balanced it to have around 200,000 samples in each class. This preprocessing step allows us to avoid any potential issue related to class imbalance such as bias toward a particular class. Other usual techniques are also utilized, we remove duplicate records and those containing undefined values.

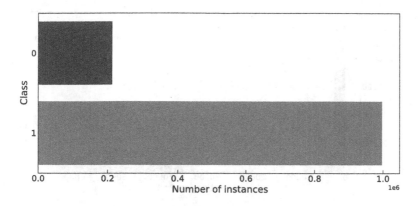

**Fig. 2.** Distribution of Classes

After this, we quickly researched for potential indicators of a class. First, we analyzed source ports used in Fig. 3. As shown, malicious traffic usually uses ports over 30,000 while benign traffic mostly uses the usual system ones as source ports such as SSH, HTTPS, FTP, etc. On the contrary, there is a clear pattern when we look at destination ports in Fig. 4. Attacks are targeting well-known ports. Unsurprisingly, in Fig. 5, we can see that HTTP is the most targeted service followed by HTTPS, DNS, and port 8080 which is often associated with web services or proxies.

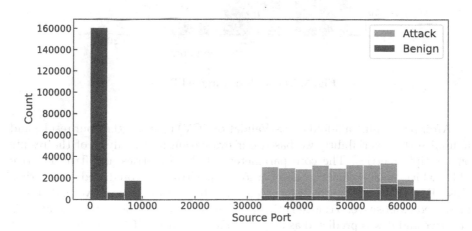

**Fig. 3.** Distribution of Source Ports Used

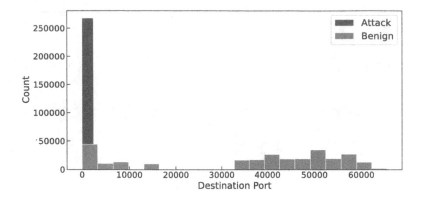

**Fig. 4.** Distribution of Destination Ports Used

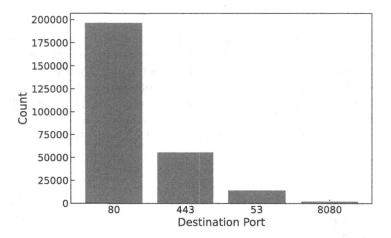

**Fig. 5.** Top 4 Most Targeted Ports

After performing a 5-fold Cross-Validation (CV) to strengthen our results and minimize their variability, we based our evaluation and analysis of the results on multiple metrics. The core parameters of these metrics are: True Positive (TP) which means that the observation is positive and predicted as positive, False Positive (FP) which means that the observation is negative and it was predicted as positive, True Negative (TN) which means that the observation is negative and it was predicted as negative, False Negative (FN) which means that the observation is positive and it was predicted as negative; accuracy, precision, recall, f1-score, Receiving Operator Characteristic (ROC) curve and Area Under the Curve (AUC).

Accuracy represents the ratio of the correct predictions (TP+TN) to the total data. this gives the following equation:

$$Accuracy = (TP + TN)/(TP + FP + TN + FN) \qquad (4)$$

Recall represents the ability of the model to identify all positive cases, given correctly:

$$Recall = TP/(TP + FN) \tag{5}$$

Precision calculates the percentage of real positive instances (properly predicted positives) out of all the occurrences the model predicts as positive. This gives the following equation:

$$Precision = TP/(TP + FP) \tag{6}$$

F1-score is the harmonic mean of precision and recall, given by:

$$F1 = 2 * Precision * Recall/(Precision + Recall) \tag{7}$$

A binary classification model's performance is also assessed using the ROC curve, which gauges how well it can discriminate between positive and negative classifications. The True Positive Rate (TPR) and False Positive Rate (FPR), which represent the proportion of positive instances correctly categorized as positive and the proportion of negative instances wrongly labeled as positive, respectively, are plotted on the ROC curve for various classification thresholds. The AUC, which has a value between 0 and 1, is a frequently used indicator of the model's overall performance. A higher AUC denotes better model performance.

**Table 1.** Performance Metrics Comparison of our ML Models with 5-fold CV

| Model | Fit Time (s) | Accuracy | Precision | Recall | F1 |
|-------|--------------|----------|-----------|--------|------|
| RF | 0.985 | 0.991 | 0.983 | 0.999 | 0.991 |
| DT | 0.510 | 0.990 | 0.990 | 0.982 | 0.990 |
| MLP | 25.04 | 0.975 | 0.970 | 0.981 | 0.975 |
| LR | 1.132 | 0.911 | 0.855 | 0.989 | 0.917 |

Given the high classification accuracies observed in our study, as summarized in Table 1, it is challenging to definitively determine the single best-performing classifier. Nevertheless, our analysis provides valuable insights into the strengths and weaknesses of each classifier.

As shown in Table 1, all four classifiers achieved remarkable classification results, with accuracy scores consistently exceeding 90%. Notably, the RF classifier emerged as a strong performer, with an almost perfect recall of 0.999. This high recall indicates the RF model's ability to effectively identify true positive cases, minimizing the risk of false negatives. On the other hand, the DT classifier achieved a higher precision score, with a value of 0.990. High precision suggests that DT excels in correctly classifying positive cases, reducing the number of false positives. The choice between recall and precision may depend on the specific needs of an IoT security application.

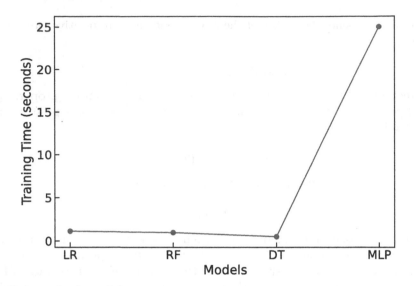

**Fig. 6.** Training Times for each Model

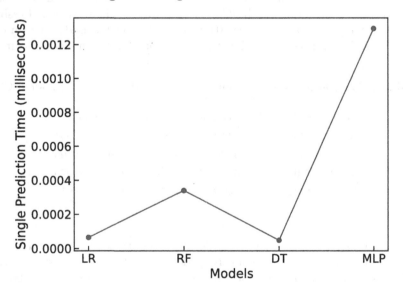

**Fig. 7.** Prediction Times for a Single Instance for each Model

In addition to classification accuracy, it is essential to consider the computational efficiency of each classifier, especially in resource-constrained IoT environments. Both RF and DT models demonstrated exceptional efficiency. They completed the training phase in less than one second, making them well-suited for real-time applications. Conversely, MLP exhibited the longest training time, requiring 25 s to complete. Moreover, MLP had a prediction time approximately three times longer than RF and DT, as depicted in Figs. 6 and 7.

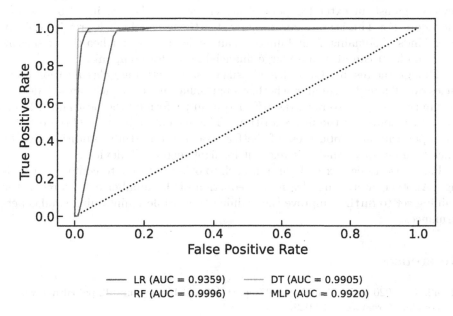

**Fig. 8.** ROC Curves of our Models

The analysis of the ROC curve, visualized in Fig. 8, provides further information into classifier performance. The ROC curve of RF indicates near-flawless performance, as demonstrated by its high AUC score of 0.9996. This suggests that RF can effectively distinguish between positive and negative cases across various threshold values, making it a strong candidate for reliable attack detection. While MLP and LR displayed lower AUC scores compared to RF, they still achieved commendable results. However, RF and DT seem to be better options in this context, according to our research.

Two factors that may have contributed to the successful classification results are the meticulous feature engineering, providing fine-grained information for the classifiers, and the normalization preprocessing, ensuring that features are on a consistent scale. These practices are essential for improving the discriminatory power of the classifiers and optimizing their ability to detect attacks in IoT devices. In summary, our results demonstrate the effectiveness of various ML classifiers in enhancing IoT device security. The choice of the most suitable classifier should depend on the specific application requirements, considering factors such as classification performance, computational efficiency, and interoperability. Our study found that RF and DT effectively address these issues.

## 5   Conclusion

Our paper aims to demonstrate the possibility of enhancing the security of IoT devices against an extended pool of attacks such as DDoS, injections, scans, backdoors, etc. This is extremely important as the usage of IoT is growing quickly in all kinds of domains. A malfunction can be destructive for healthcare sensors for example. It can also cause large financial losses for companies.

Though the results, for all five classifiers, show very high classification accuracies over 0.9 in determining whether a particular attack attacks an IoT device, we can say that, on average, the RF algorithm performed the best and LR performed the lowest of the five algorithms. The extremely high AUC score shows the superiority and robustness of the RF model in our study. Additionally, the quick training speed makes it relevant for lightweight IoT devices.

Future works can extend our approach to other datasets to confirm our findings. Another interesting idea is the reduction of the number of features or the training set to further improve the training time while maintaining similar performances.

## References

1. 42, U.: 2020 unit 42 IoT threat report (2020). https://unit42.paloaltonetworks.com/iot-threat-report-2020/
2. Abdul-Ghani, H.A., Konstantas, D., Mahyoub, M.: A comprehensive IoT attacks survey based on a building-blocked reference model. Int. J. Adv. Comput. Sci. Appl. **9**(3), 355–373 (2018)
3. Alsaedi, A., Moustafa, N., Tari, Z., Mahmood, A., Anwar, A.: ToN_IoT telemetry dataset: a new generation dataset of IoT and IIoT for data-driven intrusion detection systems. IEEE Access **8**, 165130–165150 (2020)
4. Aubet, F., Pahl, M.: DS2OS traffic traces (2018)
5. Bala, R., Nagpal, R.: A review on KDD Cup99 and NSL NSL-KDD dataset. Int. J. Adv. Res. Comput. Sci. **10**(2) (2019)
6. Booij, T.M., Chiscop, I., Meeuwissen, E., Moustafa, N., Den Hartog, F.T.: ToN_IoT: the role of heterogeneity and the need for standardization of features and attack types in IoT network intrusion data sets. IEEE Internet Things J. **9**(1), 485–496 (2021)
7. Garcia, S., Parmisano, A., Erquiaga, M.J.: IoT-23: a labeled dataset with malicious and benign IoT network traffic (2020). https://doi.org/10.5281/zenodo.4743746
8. Kang, H., Ahn, D.H., Lee, G.M., Yoo, J.D., Park, K.H., Kim, H.K.: IoT network intrusion dataset (2019). https://doi.org/10.21227/q70p-q449, https://dx.doi.org/10.21227/q70p-q449
9. Koroniotis, N., Moustafa, N., Sitnikova, E., Turnbull, B.: Towards the development of realistic botnet dataset in the internet of things for network forensic analytics: Bot-IoT dataset. Futur. Gener. Comput. Syst. **100**, 779–796 (2019)
10. Maguluri, L.P., Ananth, J., Hariram, S., Geetha, C., Bhaskar, A., Boopathi, S.: Smart vehicle-emissions monitoring system using internet of things (IoT). In: Handbook of Research on Safe Disposal Methods of Municipal Solid Wastes for a Sustainable Environment, pp. 191–211. IGI Global (2023)

11. Mahmood, A., et al.: Industrial IoT in 5G-and-beyond networks: vision, architecture, and design trends. IEEE Trans. Industr. Inf. **18**(6), 4122–4137 (2021)
12. Meidan, Y., et al.: N-BaIoT-network-based detection of IoT botnet attacks using deep autoencoders. IEEE Pervasive Comput. **17**(3), 12–22 (2018)
13. Moustafa, N., Keshky, M., Debiez, E., Janicke, H.: Federated ToN_IoT windows datasets for evaluating AI-based security applications. In: 2020 IEEE 19th International Conference on Trust, Security and Privacy in Computing and Communications (TrustCom), pp. 848–855. IEEE (2020)
14. Moustafa, N., Slay, J.: UNSW-NB15: a comprehensive data set for network intrusion detection systems (UNSW-NB15 network data set). In: 2015 Military Communications and Information Systems Conference (MilCIS), pp. 1–6. IEEE (2015)
15. Pahl, M.O., Aubet, F.X.: All eyes on you: Distributed multi-dimensional IoT microservice anomaly detection. In: 2018 14th International Conference on Network and Service Management (CNSM), pp. 72–80. IEEE (2018)
16. Sharafaldin, I., Lashkari, A.H., Ghorbani, A.A.: Toward generating a new intrusion detection dataset and intrusion traffic characterization. ICISSp **1**, 108–116 (2018)
17. Tan, S., Guerrero, J.M., Xie, P., Han, R., Vasquez, J.C.: Brief survey on attack detection methods for cyber-physical systems. IEEE Syst. J. **14**(4), 5329–5339 (2020)
18. Taylor, M., Reilly, D., Wren, C.: Internet of things support for marketing activities. J. Strateg. Mark. **28**(2), 149–160 (2020)
19. Velasco-Mata, J., González-Castro, V., Fernández, E.F., Alegre, E.: Efficient detection of botnet traffic by features selection and decision trees. IEEE Access **9**, 120567–120579 (2021)

# Attentive, Permutation Invariant, One-Shot Node-Conditioned Graph Generation for Wireless Networks Topology Optimization

Félix Marcoccia[1,2,4]([✉]), Cédric Adjih[3]([✉]), and Paul Mühlethaler[1]([✉])

[1] INRIA Paris, Paris, France
{felix.marcoccia,paul.muhlethaler}@inria.fr
[2] Thales SIX, Gennevilliers, France
[3] INRIA Saclay, Palaiseau, France
cedric.adjih@inria.fr
[4] Sorbonne Université, Paris, France

**Abstract.** It is common knowledge that using directional antennas is often mandatory for Multi-hop ad-hoc wireless networks to provide satisfying quality of service, especially when dealing with an important number of communication nodes [1]. As opposed to their omnidirectional counterpart, directional antennas allow for much more manageable interference patterns: a receiving antenna is not necessarily interfered by nearby emitting antennas as long as this receiving antenna is not directed towards these undesired emission beams. Two nodes then need to steer one of their antennas in the direction of the other node in order to create a network communication link. These two users will then be able to, in turn, emit and receive to and from each other. The scope of this work resides in finding a centralized algorithm to governate these antenna steering decisions for all users to instantaneously provide a valid set of communication links at any time given the positions of each user. The problem that raises is then a geometrical one that implies finding topologies of network links that present satisfying throughput and overall QoS and guarantee instantaneous connectedness i.e. the computed set of links allows any user to reach any other user in a certain number of hops. Building such optimized link topologies makes further tasks, such as routing and scheduling of the network, much simpler and faster. This problem is highly combinatorial and, while it is solvable with traditional Mixed Integer Programming (MIP), it is quite challenging to carry it out in real time. For this purpose, we propose a Deep Neural Network that is trained to imitate valid, solved instances of the problem. We use the Attention mechanism [2,3] to let nodes exchange information in order to capture interesting patterns and properties that then enable the neural network to generate valid network link topologies, even dealing with unseen sets of users positions.

É. Renault et al. (Eds.): MLN 2023, LNCS 14525, pp. 14–31, 2024.
https://doi.org/10.1007/978-3-031-59933-0_2

# 1    Introduction

Multi-hop wireless networks, especially with the current deployment of 5G technologies and research work on 6G, constitute a very active field of research. While
the per-user capacity and throughput of a network are known to scale poorly
with the increase of users [4], it has been proven that using directional antennas with reduced beamwidth can help mitigate the loss in Quality of Service
(QoS) by a factor inversely proportional to the beamwidth [1]. Using directional
antennas allows for both higher per-link throughput and much better interference management. One must chose wisely, for each antenna, towards which node
to point. Indeed, we want to steer all the antennas in a way that avoids creating low SNR links and high interference patterns. In order for the antennas
to be aligned at transmission time, paths from users to others have to be computed in advance, either in the form of routes computed along with the resource
allocation and traffic control level (OSI layer 3), or even beforehand to reduce
later computations, by computing carefully an optimized network topology (OSI
layer 2) to allow for easier network and traffic management afterwards. Usually,
on top of the network constituted by the links, proper routing and scheduling
need to be computed. An example is given by [5], that assumes a slotted frame
structure as well as some known antenna orientations, and that, on that basis,
computes a routing table and then establishes a transmission schedule. Some
important network properties have to be guaranteed, the most crucial one being
connectedness. Moreover, the network must respect several physical constraints.
We observe that this problem has a highly combinatorial aspect: global connectedness relies on complex combinations of links, and physical constraints require
taking into account interdependence and interference between created links. We
aim to create a topology that provides instantaneous global connectedness and
fair adherence to physical constraints to alleviate further network tasks. In an
ideal scenario, where the topology is optimal, routing can then be simple shortest
paths and scheduling can consist in simple 2-slot scheduling where one half of
the nodes emits while the other half receives and vice versa. This is of course a
complex problem to tackle. To mitigate some of the combinatorial and computational burdens, we propose a one-step process that generates such a topology. We
train a Deep Neural Network to imitate the results of an Integer Programming
(IP) instance of the problem. It can then be used to infer graphs that hold the
same properties as the ones labeled as solutions of the IP problem, in constant
time.

Graphs are heterogeneous objects, composed of both nodes and edges, which
do not have any natural ordering. They then require an Neural Network architecture that can feature both node and edge computations and respect some
crucial properties not to bring undesirable biases due to the nodes' or the edges'
ordering. This is the main focus of our work. The rest of the article is organized
as follows: in Sect. 2, we present articles and literature related to optimizations
of such networks. In Sect. 3, we detail the system model and our problem state-
ment. Then we introduce our solution and its properties in Sect. 4.1. In Sect.

6, we present some results that confirm the value of our method and illustrate how its components impact its performance, thus proving their merit. We finally conclude in Sect. 7.

## 2 Related Work

Recently, there has been a significant push to develop more effective network solutions, particularly with the emergence of 5G and 6G technologies. Integer Programming is commonly employed for finding optimal solutions, often utilizing linear formulations for paths, links, and flows as seen in sources like [6]. However, to quicken this process, different algorithms and heuristics are frequently implemented, as in [5,7]. Despite speeding up solution computation, these methods are usually iterative and rely heavily on greedy algorithms, which can be a drawback in certain practical applications.

Deep Learning offers a notable advantage in this context. It simplifies the solution generation process to a sequence of matrix multiplications using pretrained coefficients. This approach effectively transforms combinatorial challenges into nonlinear, multivariate statistical problems with parametric nuances. Deep Learning is particularly beneficial because it scales efficiently with user numbers, maintaining linear complexity relative to the input nodes - a stark contrast to the scalability issues faced by combinatorial methods with numerous input nodes.

Deep Neural Networks (DNNs) are versatile and can be applied to both simple tasks like network performance prediction (as in [8]) and more complex ones like inferring network graphs, often through dynamic temporal prediction methods (as in [9]). The field of Deep Learning for graph structures, including Graph Neural Networks (GNN) [10,11], has been gaining immense attention. Generative graph models like GraphVAE [12], which use a GNN and a global pooling operation to represent entire graphs, facilitate continuous and potentially conditioned graph generation. GNNs typically offer edge-conditioned convolution or message passing techniques. In our case, without actual edges to analyze, we employ an alternative feature extractor aligned with some of the main GNN principles that ensure permutation invariance. Here, the well-known Attention mechanism, first introduced in [2] and popularized through its use in the Transformer model as demonstrated in [3], becomes highly relevant. This mechanism offers a promising approach, akin to a retrieval system, and its effectiveness has been empirically validated. Attention is generally popular to treat graph-shaped data, as [13,14]. Attention could also benefit from some of the expressive power proved in [15,16] due to its Query/Key Retrieval formulation that implies some form of successive node-matrix multiplications in the layers. As a solution for node-conditioned network topology generation, nodes2net [17] was proposed, the authors use Attention and a flattened linear layer to get link predictions. While it gives good results, the generalization lacks theoretical guarantees and can be inconsistent because of the flat linear layer, which can be greatly biased by the ordering of the nodes. The fixed-sized linear layer also implies that the model is not flexible regarding the number of communication nodes.

# 3   System Model

The system we wish to study is the same as the one in [17]. We consider a set $V$ of $n$ nodes described by their respective 2D coordinates $(x, y)$. We assume an idealized "protocol model", where nodes can transmit if they are within the radio range. Each node has a fixed number of independent antennas, each of which can establish at most one link. Candidate links correspond to pairs of nodes that are in range of each other, and hence are represented as undirected edges. Our problem is to find a subset of those edges, denoted $E$, that satisfies some constraints (antenna number, placement, more importantly connectedness...) while possibly optimizing some objective function $f$. $E$ amounts to the links obtained after orienting (configuring) antennas. Once the edges are given, we then have a graph $G = (V, E)$ that can serve as the network link topology. As a comprehensive example, and throughout this work, we will consider the problem of the creation of a network of $n$ nodes under some physical link constraints (limited number of communication links per user, a fixed maximum length for each link...) that must ensure global connectedness while minimizing the total number of links. It can be viewed as an optimization problem with several constraints corresponding to physical limitations of the communication links. We formalize our problem as an Integer Programming one to obtain optimal solutions, and will train a neural network to output graphs as close as possible to these solutions.

Our problem can be formalized as follows:

- $V$ is the set of nodes 1, 2, ..., n
- $e_{i,j}$ is a binary decision variable that indicates that there is an edge between node $i$ and node $j$ (i.e. after orientation of one of their antennas to create this link)
- One node is also described by its 2D coordinates $x_i, y_i$

We want to solve the following optimization problem

$$\min_{e_{i,j}} \sum_{i=1}^{n} \sum_{j=1}^{n} e_{i,j} \tag{1}$$

s.t.

Logical and physical constraints (2)–(9):
One node holds $n_{\text{antennas}}$. Since one antenna can not form several links one node can then form at most $N_{\text{antennas}}$ links:

$$\forall i \in V, \quad \sum_{j=1}^{n} e_{i,j} \leq N_{\text{antennas}} \tag{2}$$

One link can be formed between two nodes only if they are within a maximum radio range $D_{max}$ one from the other:

$$\forall i, j \in V^2, \quad e_{ij} \times [(x_j - x_i)^2 + (y_j - y_i)^2] \leq D_{max} \tag{3}$$

Links are bidirectional:

$$\forall i, j \in V^2, \quad e_{i,j} = e_{j,i} \tag{4}$$

One node can not form a link with itself:

$$\forall i, j \in V^2, \quad i = j \Rightarrow e_{i,j} = 0 \tag{5}$$

In addition, it is necessary to establish a mathematical formalization for the connectedness of the entire network. A commonly used technique is to introduce phantom flows originating from a virtual source at a specific node (without loss of generality, let's denote it as $v_0$). These phantom flows are required to traverse through every node in the network, and flow conservation equations are formulated accordingly. The network is considered connected if such flows can be identified. The constraints that enforce connectedness through phantom flows are as follows:

Virtual source distributes $n - 1$ flows through the network:

$$\sum_{j=0}^{n} f_{0,j} = n - 1 \tag{6}$$

Flows propagate through existing links:

$$\forall i, j \in V^2, \quad i \neq j \Rightarrow f_{i,j} \leq (n - 1) \times e_{i,j} \tag{7}$$

Each node absorbs one flow and transmits the rest:

$$\forall i \in V, \quad \sum_{j=0}^{n} f_{i,j} = \sum_{j=0}^{n} f_{j,i} - 1 \tag{8}$$

One node can not send a flow to itself:

$$\forall i, j \in V^2, \quad i = j \Rightarrow f_{i,j} = 0 \tag{9}$$

## 4   Our Approach

We previously mentioned that carrying out such Integer Programming problems in real time can be difficult, especially with an important number of communication nodes. This is why our goal is to be able to extract the essence of the distribution of the solution graphs with a monolithic structure that can then generate new valid graphs when dealing with new, unseen sets of nodes.

### 4.1   Model

In this paper, we propose using Deep Neural Networks to infer valid sets of links, using the nodes' positions as inputs, by "imitating" optimal instances obtained by solving the aforementioned linear program (Fig. 1).

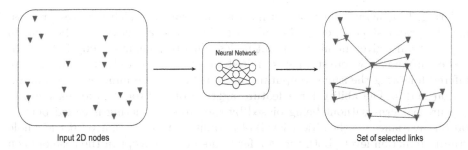

Input 2D nodes                                                           Set of selected links

**Fig. 1.** Overview of our solution: the model takes the nodes' positions as input and outputs the set of edges to connect them. The 2D positions of the nodes and the selected links correspond on one actual dataset sample and its proposed solution.

Our model takes as inputs the set $V$ of nodes described as:

$$v_i = (x_i, y_i) \quad \forall i \in [1, n].$$

It outputs an adjacency matrix:

$$E_{pred} = \begin{pmatrix} e_{11} & e_{12} & \cdots & e_{1n} \\ e_{21} & e_{22} & \cdots & e_{2n} \\ \vdots & \vdots & \ddots & \vdots \\ e_{n1} & e_{n2} & \cdots & e_{nn} \end{pmatrix}$$

that describes the set $E$ of edges of the desired graph.

We want to find a function F, namely a Deep Neural Network, parametrized by its weights $\theta$, such that

$$F_\theta(V) = E_{pred}.$$

In the training phase, we first solve the optimization problem (1)–(9) on some random instances of graphs through a MILP solver (after linearization of the equations). We use these solved instances to constitute the dataset mapping input nodes' positions to their respective desired topology of links. For each datapoint of nodes' positions $V$ this hence yields the label adjacency matrix corresponding to the true solution called $E_{pred}$, that the model should then "predict". We then want to find the best model to learn this mapping between nodes and topologies.

**Permutation Invariance and Attention**
While a Multi-Layer Perceptron (MLP) may initially seem like a straightforward network architecture well suited for such problems, it is biased by the ordering of the nodes. One main issue with creating a graph from nodes is indeed the need for a feature aggregator that does not process nodes as an ordered sequence, which would for instance result in a neuron learning some combination of input 1 with input 2 and 4, which would not make sense and would obviously lead to undesirable bias since, as stated before, nodes do not feature any natural

ordering. In general, achieving permutation invariance, which involves processing the graph at node-level, also allows a single neural network to be able to work with any given number of input nodes. This particularly useful for embedded environments where storage is limited, and to make sure that topologies are infered following the same computations regardless of the number of communication nodes. The ability for a feature aggregator to process information and capture patterns without being biased by the objects ordering is called permutation invariance (at the object level) of permutation equivariance (at the whole graph prediction level), both terms refer to a similar concept in the case of deep learning on graphs. While GNNs use edge-conditioned, spectrally defined convolutions, or strict message passing algorithms, most of them are not perfectly suited for a whole graph prediction task since the edges are yet to be predicted, and hence do not hold information that can help the prediction process. The global guideline a permutation invariant aggregator has to follow is to process at node and possibly edge-level, in order to capture some underlying local distribution that explains the link patterns of the dataset's graphs. Through the layers, the feature aggregation process should be able to capture larger patterns, involving more nodes and broader neighbourhoods of nodes, making the patterns somewhat "global" while keeping the computation at node and edge-level.

We use the Attention mechanism [2, 3] as the message-passing modality of our model. It enables nodes to exchange information in a way that allows fine feature aggregation amongst several nodes without breaking permutation invariance. We also use residual connections [18] from the input to further layers not to lose the position signal of each node and between layers not to lose each node's individual embedding. The most popular version of Attention is the one used for Transformer models [3], The simplified embeddings of the set of nodes $V$ going through one Attention layer $l$ (for one Attention head, in practice we use up to 8) can be described as below:

$$H_{l+1} = \text{softmax}\left(\frac{W_q H_l (W_k H_l)^T}{\sqrt{d_k}}\right) W_v H_l \oplus H_l \oplus H_{l-1}$$

where $H_0$ represents the initial features of the inputs nodes, $H_l$ the embeddings of the nodes at layer l and $\oplus$ denotes the residual connections (we add previous layers signals and apply normalization over the output). $W_q, W_k$ and $W_v$ are learnable matrices that are what we train to form the Queries, the Keys and the Values, in order to get the Attention to capture the relevant features. We use MultiHead Attention, which defines several different instances of these learnable weights to allow attending to different characteristics, then aggregates them.

The idea is that one node $v_j$ which is relevant to attend to for a node $v_i$ will learn to define a $W_k K_j$ rather similar (to maximize $W_q Q_i W_k K_j$) to $W_q Q_i$ such that $v_j$ "wins" the softmax i.e. $v_i$ "understands" that it must attend to node $v_j$. The point is then to learn a parametrization of the weight matrices $W_q, W_k$ in order to learn a reliable transformation that captures relevant pairs of nodes so that they present well-aligned Query and Key. The Value weight matrix $W_v$ learns how to finally embed the node combining the attended nodes information.

We use Multihead Attention as it is naturally fitted to treat unordered tokens, and provides several good properties: it allows nodes to exchange information based on different criteria without necessarily needing edges as trustable mean of propagation, it can greatly avoid oversquashing (where long range dependencies tend to vanish as the number of message passing intermediates augment) and it can learn generalizable and intelligent gatherings of nodes' coordinates to create systematic patterns. Attention can also be prone to oversmoothing, particularly inside dense neighbourhoods of nodes, where the attention scores tend to explode and lead the nodes' embeddings to become similar, or get one node to be attended by or attending to an abnormally high number of other nodes, implying poor and altered embedding. Using skip connections can be a way to deal prevent some of these issues, as we will detail further on (Fig. 2).

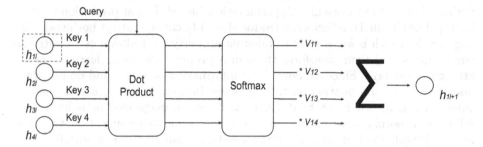

**Fig. 2.** Illustration of one Attention embedding layer applied to one node $v_1$ in a 4-node-network

While this Attention-based message-passing scheme is appropriate, one major challenge remains: how can we infer links from these node representations ?

## 5   Our Attention-Based Solutions

### 5.1   Attention-Only Model

One model we implement is an Neural Network that uses Attention both as the feature aggregator and for the prediction layer. Nodes exchange through Attention-based message passing as described in Fig. 3, and infer their predictions in the form of Attention scores. The prediction part of the network then only consists in such Attention scores computed by each node for each other node to go through unit-sized linear layers in order to output 0 s or 1 s, or to be directly used as thresholded predictions. The computation of the attention scores, can then be formulated as:

$$E_{pred} = \text{softmax}\left(\frac{W_q H_l (W_k H_l)^T}{\sqrt{d_k}}\right)$$

We can also use unnormalized attention scores, which loses some of the benefits and "logistic" expressivity of the softmax, but allows greater range of values for link predictions, in order to get 0 s and 1 s more easily. In that case the formulation of the link predictions is simply:

$$E_{pred} = \frac{W_q H_l (W_k H_l)^T}{\sqrt{d_k}}$$

In practice, we use the softmaxed values as it allows better global coherence of the link predictions and generally create more plausible sets of links.

## 5.2   Graph Transformer

We hereby present a permutation-invariant structure that allows to infer a linkset without having to utilize a global, permutation-biased, Linear prediction layer as in [17]. The Graph Transformer, introduced in [14], can operate at both node and edge levels, which is an incredibly valuable property as it allows us to initialize edge objects and then transform them into proper links using learned "node-attentive" patterns. Since we don't have information on edges and do not want node-redundant information on the edges, we initialize them as simple neutral values (ones, in order not to disturb fist layers propagation) or noisy values following a normal distribution. In case where some communication links are impossible (physical obstacle, discretion needs, jamming...) we can initialize their value at 0, so that the model processes it as an impossible link. [19] to improve graph-level feature extraction. We add skip connections between layers, in order to stabilize gradients and help access previous layers' information.

The principle of the model is to rely on the Attention computed between nodes to modulate the edge features. These edge features are also used to modulate the Attention-based signal propagation between nodes. Nodes and Edges are then updated interdependently and nodes serve as intermediates to create plausible edge patterns. We use residual connections, that prove to be important for deeper Neural Networks, and once again equip the model with Registers. We do not really use Positional Embeddings, that usually are materialized by Laplacian Eigenvectors [20,21] since we generally initialize our edges with ones and, even with random values, the positional encodings did not appear to make significant difference in the results, for a slightly more important computational cost due to the eigendecomposition.

The simplified embeddings of the set of nodes $V$ going through one Graph Transformer layer $l$ (for one Attention head, in practice we use up to 8) can be described as below:

$$H_{l+1} = \text{softmax}\left(\frac{W_q H_l (W_k H_l)^T}{\sqrt{d_k}} . W_e E_l\right) W_v H_l \oplus H_l \oplus H_{l-1}$$

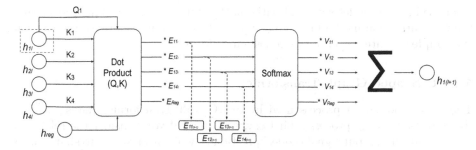

**Fig. 3.** Illustration of one Graph Transformer (dotprod attention) layer applied to one node $v_1$ in a 4-node-network

and

$$E_{l+1} = \frac{W_q H_l (W_k H_l)^T}{\sqrt{d_k}} . W_e E_l$$

with $E_0$ being a linear projection of the initial edge embeddings

We also implement a model which presents a similar structure but whose Attention is additive, as in [2]. In this case, the Attention between two nodes is computed by a 1-layer neural network that is fed the two nodes embeddings. We implement it in order to study whether it can help avoid some of the typical oversmoothing that appear in graph processing tasks, that the similarity-based retrieval formulation used by Dot-Product Attention does not totally avoid.

## 5.3   Disentangled Transformer

We propose a model whose the edge embeddings are a bit disentangled from the Attention computed between nodes. Here for an edge between $i$ and $j$ a small MLP is fed the concatenation of the $i$ and $j$'s embeddings to infer a edge-level embedding.

$$E_{ij} = tanh \left( W_e \begin{bmatrix} h_i \\ h_j \end{bmatrix} + b_e \right)$$

It then modulates the attention scores of the layers as in the standard Graph Transformer. We implement this component between Attention computation and the final edge embeddings, in order to add some flexibility to the edge embedding process. It is meant to alleviate some of the Attention tasks, since it has both to operate some information gathering message passing and to arrange Attention scores to form plausible edges. Here both are somewhat disentangled even though they are of course interdependant.

We also implemented a model that would only process the concatenated pairs of nodes and use Attention between the pairs, but it did not perform well as the predictions of the edges were almost totally independent, which means any rather short edge would be predicted true with little to no respect to the other predicted edges or nodes. Indeed, the model did not seem to make great use of Attention, as we could observe when displaying the Attention scores and when trying the model with and without the Attention layer. This proves that the node-attention method to get edge predictions is much more expressive that the simple concatenation of the nodes' embeddings.

### 5.4   U-Shaped Graph Transformer

The edge modulation process used by our Graph Transformer tends to make the node embedding process a bit lossy, especially if you consider the relatively simple and small input signal (nodes' positions): while it allows us to shape edges in a pleasant way, after several layers, the nodes can suffer from oversmoothing, or can have "derived" a bit. The learned edges are mostly "taught" to form probable links, and hence might not necessarily be perfectly suited for message passing, especially in the last layers, which are the results of many successive edge-modulated message passing. To address this issue, we connect last layers to the first ones in a U-Net [22] way. It is important to state that the model does not feature an downsampling-upsampling process, as U-Nets do (notably because graph upsampling is much more difficult than in an image processing context), the U-shape only comes from the use of skip connections between first and last layers as shown in Fig. 4. It enables the network to access once again the nodes' positions but, this time, with a almost formed set of links that bring a desirable bias that can guide the final prediction of the edges. We also implement it with the Attention-only model but it does not seem to be as relevant and important given the model size and the absence of edge modulation. For instance, the value of the nodes of the last layer would be:

$$H_{last} = \text{softmax}\left(\frac{W_q H_l (W_k H_l)^T}{\sqrt{d_k}}.W_e E_l\right) W_v H_l \oplus H_{last-1} \oplus H_0$$

### 5.5   Registers

As these edges are only modified via node-level Attention schemes, even if these can benefit from great expressivity and receptive field thanks to their consecutive node-to-node signal exchange, we wish to add a component aimed at improving global consistency by allowing to "store" graph-level attributes and patterns. We follow the guidelines established in [19] and adapt them to graphs. We add one or several dummy nodes whose initial embedding values are learned. These dummy nodes can then serve as registers, meaning that, through Attention-based message passing, they can be attended by the nodes of the graph to both store and retrieve information. It is useful to gather information that might not directly be related or useful to a single node or neighbourhood of nodes but, combined with many information signals from different nodes of the graphs it can aggregate in these registers as graph-level features. This type of global, long dependency gathering of information is generally a problem in graph message passing algorithms as nodes tend to keep neighbourhood-level interactions, as it suffers from oversquashing.

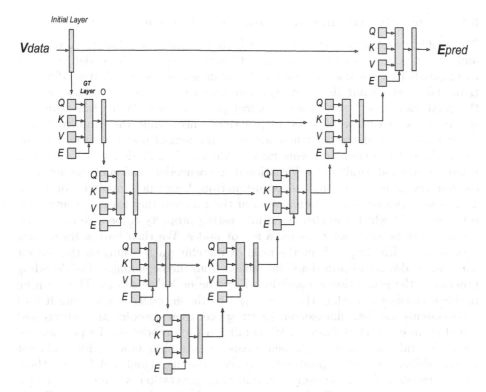

**Fig. 4.** Illustration of a simplified U-shaped Graph Transformer, first and last layers are directly connected

## 5.6 Graph-Level Attributes

We can also add graph-level attributes by adding them as inputs and concatenating them with the nodes, allowing the nodes to attend to them (Register or Cross-Attention), or using FiLM Layers [23,24] or a HyperNetwork [25] architecture. We then search $F$ parametrized by $\theta$ such that

$$F_\theta(V, \gamma) = E_{pred}$$

where $\gamma$ is a sequence of graph level attributes. Graph attributes can be for example several spectral properties such as the algebraic connectivity, or can be related to the density of links, centrality, and can also be learned. In the case where the attribute can be directly expressed as a differentiable function of the output of the model and its weights, the respect of the desired the attribute in the predicted matrix can be enforced using Lagrangian Duality framework described in [26]. For instance, one could use an approximated differentiable expression of the connectedness or the total number of links or such, from the output of the model, and enforce it to be as close as their chosen value.

## 5.7    Generative Formulation: Variational Inference

The problem itself is not deterministic, which might lead to sub-optimal training and results in the supervised learning setting: the output of the model, even if it corresponds to a possible solution for the given set of nodes, might be different than the exact output that was expected, and cause large loss values, even if the prediction was "valid" from a logical point of view. While this issue has not been extremely problematic in practice, since, while the problem is not deterministic, the dataset is (there are no equal sets of nodes' positions in the dataset), and the training seems rather stable. We still derive a Variational Inference method, similar to a Variational Autoencoder version of the network, so that the problem becomes a reconstruction, hence deterministic, problem. This is a proper generative formulation of the problem that learns a continuous solution space, which can also be an interesting property to be able to explore the different possible solutions for a set of nodes. We then have a framework comparable with [12], without the need of matching algorithms on the output since our nodes are identified and initialized from the beginning of the decoding process as the generation is conditioned by the nodes' positions. The learning problem consists in feeding the true graph to the encoder, compressing it into a continuous and low dimensional latent space, then decoding the compressed signal to reconstruct the inputs. We condition the decoder on the positions of the nodes. Infering a new graph then consists in sampling in a continuous latent space, adding the nodes' positions as a conditioning signal and feeding them to the decoder. To enable sampling, and thus generation, we want a smooth, continuous latent space. To do so, we enforce the encoder's output to follow a multivariate Gaussian distribution. On the other hand, we train the decoder to output values close to the input being reconstructed. The encoder consists in a graph feature aggregator, typically GNN, followed by pooling layers in order to obtain a graph level encoded vector $z$. The decoder can be any of the previously described models (Attention model, Graph Transformer) described above but we add the encoded latent vector by either concatenating it with each node, or as a vector to be consulted through Attention.

The problem then results in maximizing the following:

$$\mathcal{L}(\theta, \phi; \mathbf{G}) = \mathbb{E}_{q_\phi(\mathbf{z}|\mathbf{G})} \left[\log p_\theta(\mathbf{E}|\mathbf{V}, \mathbf{z})\right] - D_{KL}\left(q_\phi(\mathbf{z}|\mathbf{G}) \| p(\mathbf{z})\right)$$

$q_\phi$ represents the approximated posterior latent distribution
$p_\theta$ represents the approximated likelihood of the data
$p(z)$ represents the prior distribution of the latents, we assume it to follow a multivariate Gaussian distribution.

The first term is the Reconstruction Loss, maximizing it enforces the reconstructed edges to be as close as possible to the input edges. The second term is the Kullback-Leibler Divergence, minimizing it enforces the approximated latent distribution to be close to the prior distribution $z$ that is assumed to follow a multivariate Gaussian.

Infering a graph prediction hence consists in the network estimating:

$$decoder(\mathbf{V}) = p_\theta(\mathbf{E}|\mathbf{V}, \mathbf{z}).$$

The idea that motivates this formulation is that, despite the strong signal brought by the position of the nodes, the variational encoder can still struggle to find the exact link topology corresponding to this set, especially if several solutions corresponding to a similar set of nodes lie in the dataset. In this case, the encoded vector, which brings additional information about the graph to be decoded, can help distinguish the exact solution we are looking for. Since it is continuous, the latent space in which this encoded vector lies can then be parsed to produce various solutions. In our case, we observed that the model mainly learns a mapping from nodes' positions to predicted links without making great usage of the encoded vector. The training and the results did not happen to be much different than the supervised counterpart, notably because the nodes' feature space is rather large hence it is really unlikely to find similar sets of nodes in the dataset. Sampling through the latent space to feed the decoder different latent vectors for a same set of nodes still showed to produce different outputs, which means that the encoded vector is still somehow taken into account.

# 6   Experimental Results

We conduct our experiments with an Intel Xeon(R) E5-2650v3 at 2.30 GHz CPU and a Tesla T4 GPU. With such hardware, inference time does not exceed a few ms even with the largest versions of the model. Our method is implemented using PyTorch. We use AdamW optimizer with weight decay rates between 0.1 and 0.15, results below are given for models trained for an equivalent of at most 20 epochs at $1e-4$ learning rate. We generally use a standard binary cross entropy loss function. We use a dataset of multiple instances with randomly generated nodes' positions. It is composed of 180k samples of 16 nodes' positions and the adjacency matrix obtained solving the IP problem with a solver (namely Gurobi). Generating such a dataset is long and costly so we can not, for now, notably increase its size. To prevent overfitting, we wish to keep our models' number of parameters roughly in the same order of magnitude as our dataset size. Variational and supervised versions of the training and the models showed to display similar results and similar variations so we do not feature and distinguish both in the benchmarks.

We found that the Transformer-ish models sometimes tended to make conservative predictions between 0 and 0.4 instead of values closer to 0 and 1 so we added a sparsity enforcing penalty in the loss:

$penalty = \frac{|y|}{(\overline{p}-y)^2}$ with $y$ being the prediction tensor and $\overline{p}$ an empirically observed mean for non-zero predicted values, in order to push predictions away from it (in practice it was close to 0.35). The Attention-only model showed to perform better with solely two rather large Attention layers, and no activation function, but we did not get it to output proper 0 s and 1 s without significant loss in relevance of the results. It would also output values rather continuously distributed from 0 to 0.4 so a simple 0.5 threshold would not work well, we then adaptively chose the threshold that offered the best accuracy, which in general was close to 0.25. The accuracy is measured as in a classical link prediction task for each link: we round the output of the network to either 0 or 1 and we measure the difference with the true label for each entry of the adjacency matrix. The dataset of Table 1 contains almost 75% zeros, it is easier to predict the absence of link than to predict a link, so this has to be taken into account when reading the accuracy scores. We include the accuracy on a substract of the dataset (seen in training), because the task of imitating the samples is interesting in itself, to see how well the model can theoretically reproduce these graphs in the setting where they offer the best possible generalization property obtained for the test set (not seen in training or validation). We observe that models generally reproduce dataset graphs with high accuracy, Test set graphs, which the graphs have not seen during training, are not as well reproduced. This is not necessarily an issue since, as stated before, one set of nodes can admit several optimal solutions, and many more at least plausible solutions, which makes it really difficult to find the exact one that is expected. Both models still perform much better than the MLP baseline, and of course much better than random or only 0 s guessing. The Attention-only model performs surprisingly well but requires an adaptive threshold. It happened to perform much better when using two Attention layers then using the Attention scores directly as outputs. The downside is that it can not output values high enough for us to perform a simple 0.5 thesholding. Any more complex versions of the model would perform much more poorly. Overall, the best performing model is the Graph Transformer with the U-shaped skip connections. Graph Transformers globally show good performance and very few false negatives, but tend to be a bit greedy locally, predicting too many links (Table 2).

**Table 1.** Accuracy of both dataset and unseen graphs, results show that our models vasty outperform a baseline MLP and showcase the utility of some of the components.

| Model | Accuracy on dataset | Test accuracy |
| --- | --- | --- |
| Graph Transformer | 94.70% | 88.68% |
| 2-layer Graph Transformer | 93.55% | 87.48% |
| **U-shaped Graph Transformer** | **95.04%** | **88.83%** |
| U-shaped Graph Transformer (additive attention) | 94.69% | 88.79% |
| U-shaped Disentangled Graph Transformer | 94.78% | 88.67% |
| Attention-only | 94.49% | 88.59% |
| Flattened MLP (NOT permutation invariant) | 89.86% | 84.73% |

**Table 2.** Verification of the range constraint on test graphs, Graph Transformers seem to produce more reliable links, results highlight the merit of the different components.

| Model | % of links of valid length |
|---|---|
| Graph Transformer | 96.3% |
| 2-layer Graph Transformer | 94.05% |
| U-shaped Graph Transformer | 97.41% |
| U-shaped Graph Transformer (additive attention) | 97.27% |
| **U-shaped Disentangled Graph Transformer** | **97.59%** |
| Attention-only | 94.84% |
| Flattened MLP (NOT permutation invariant) | 85.37% |

We observe that, while Graph Transformers only beat the Attention model on the link prediction accuracy test by a small margin, they tend to output much more consistently valid links. We indeed observed that the Attention model could sometimes output too long hence unrealistic nodes while the Graph Transformers would rarely make such mistakes. We can deduct that the Attention-score prediction does well in terms of link patterns consistency, as the number of predicted links is more realistic, globally and per node, but that the edge embedding process of Graph Transformers allow them finer control over the edges characteristics. Once again, the U-shaped Graph Transformers perform a bit better than the original Graph Transformer, and the ones that feature small MLPs to process the pairs of nodes seem to display slightly better results.

Since the aggregation mechanism is local, between nodes, the need to obtain complex, coherent patterns that involve high level view and combination of nodes' features requires the node embeddings to benefit from a large receptive field, which typically profits from models to be rather deep. In practice, we tested up to 10 Transformer-ish layers for each Graph-Transformer-like model and chose to feature 7 of them for most experiments. We empirically notice a significant increase of performance with the increase of the number of layers, while the simple Attention score prediction model does not really benefit from it. The process of "manipulating" the edges through the layers hence probably particularly enjoys deeper networks. Unfortunately, the deeper the models are (particularly when approaching 10 layers) the smaller the layers are, since we wish to keep the number of parameters in the same order of magnitude as the dataset size, as mentioned before.

# 7   Conclusion

We tackled throughout this work the problem of infering optimized network topologies in constant time with a Deep Neural Network with strong theoretical generalization properties that is trained to imitate a dataset following a given algorithm (a Integer Programming solver in our case). We covered the notion of

Permutation invariance and how important it is when dealing with graphs. We then showed how Attention is a perfect feature aggregator for such graph problematics and proposed several versions of Attention-based models from a pure Attention model using attention scores as predictions to Graph Transformers. Some additional architecture elements to Graph Transformers, namely U-shaped skip connections and Registers have been implemented to help us tackling our problem. We obtained very satisfactory results that highlight the utility of the components of our models. We hence provide a flexible, constant time model that provides trustable topologies of network links given the nodes' positions. It can greatly help instantaneously reaching some very important structural properties (connectedness...) and feature the desirable traits of the graphs that we train the model on. We highlighted the capacities of the different proposed architectures their contribution to the expressivity of the and the robustness of the predicted links. We hope our work has highlighted the interest of using Deep Learning methods to tackle combinatorial problems by turning them into monolithic global graph generation problems. Future steps will aim at improving the reliability of the infered topologies, possibly by introducing some link pruning algorithms. While this work is purely focused on one-shot generation of graphs, we will also investigate more progressive methods such as Denoising Diffusion Probabilistic Models [27], which follow similar "global generation" baselines but in an Langevin-dynamic-ish formulation.

# References

1. Yi, S., Pei, Y., Kalyanaraman, S.: On the capacity improvement of ad hoc wireless networks using directional antennas. In: Proceedings of the 4th ACM International Symposium on Mobile Ad Hoc Networking & Computing, MobiHoc 2003, New York, NY, USA, pp. 108–116. Association for Computing Machinery (2003). https://doi.org/10.1145/778415.778429. ISBN 978-1-58113-684-5
2. Bahdanau, D., Cho, K., Bengio, Y.: Neural machine translation by jointly learning to align and translate. arXiv:1409.0473 [cs, stat] (2016)
3. Vaswani, A., et al.: Attention is all you need. arXiv:1706.03762 [cs] (2017)
4. Gupta, P., Kumar, P.R.: The capacity of wireless networks. IEEE Trans. Inf. Theory **46**(2), 388–404 (2000). https://doi.org/10.1109/18.825799, http://ieeexplore.ieee.org/document/825799/. ISSN 00189448
5. Benhamiche, A., da Silva Coelho, W., Perrot, N.: Routing and resource assignment problems in future 5G Radio access networks (2019). https://doi.org/10.5441/002/inoc.2019.17
6. Feng, W., Li, Y., Jin, D., Su, L., Chen, S.: Millimetre-wave backhaul for 5G networks: challenges and solutions. Sensors **16**(6), 892 (2016). https://doi.org/10.3390/s16060892, https://www.mdpi.com/1424-8220/16/6/892. ISSN 1424-8220
7. Bao, L., Garcia-Luna-Aceves, J.J.: Topology management in ad hoc networks. In: Proceedings of the 4th ACM International Symposium on Mobile Ad Hoc Networking & Computing, MobiHoc 2003, New York, NY, USA, pp. 129–140. Association for Computing Machinery (2003). https://doi.org/10.1145/778415.778432. ISBN 1581136846

8. Rusek, K., Suárez-Varela, J., Almasan, P., Barlet-Ros, P., Cabellos-Aparicio, A.: RouteNet: leveraging graph neural networks for network modeling and optimization in SDN. IEEE J. Sel. Areas Commun. **38**(10), 2260–2270 (2020). https://doi. org/10.1109/JSAC.2020.3000405. ISSN 1558-0008

9. Lei, K., Qin, M., Bai, B., Zhang, G., Yang, M.: GCN-GAN: a non-linear temporal link prediction model for weighted dynamic networks. arXiv:1901.09165 [cs] (2019)

10. Scarselli, F., Gori, M., Tsoi, A.C., Hagenbuchner, M., Monfardini, G.: The graph neural network model. IEEE Trans. Neural Netw. **20**(1), 61–80 (2009). https:// doi.org/10.1109/TNN.2008.2005605. ISSN 1941-0093

11. Bronstein, M.M., Bruna, J., LeCun, Y., Szlam, A., Vandergheynst, P.: Geometric deep learning: going beyond Euclidean data. IEEE Signal Process. Mag. **34**(4):18–42 (2017). https://doi.org/10.1109/MSP.2017.2693418, arXiv:1611.08097 [cs]. ISSN 1053-5888, 1558-0792

12. Simonovsky, M., Komodakis, N.: GraphVAE: towards generation of small graphs using variational autoencoders. arXiv:1802.03480 [cs] (2018)

13. Veličković, P., Cucurull, G., Casanova, A., Romero, A., Liò, P., Bengio, Y.: Graph attention networks (2018)

14. Dwivedi, V.P., Bresson, X.: A generalization of transformer networks to graphs (2021)

15. Maron, H., Ben-Hamu, H., Serviansky, H., Lipman, Y.: Provably powerful graph networks (2020)

16. Azizian, W., Lelarge, M.: Expressive power of invariant and equivariant graph neural networks. In: International Conference on Learning Representations (2021). https://openreview.net/forum?id=lxHgXYN4bwl

17. Marcoccia, F., Adjih, C., Mühlethaler, P.: A deep learning approach to topology configuration in multi-hop wireless networks with directional antennas: nodes2net. In: 2023 12th IFIP/IEEE International Conference on Performance Evaluation and Modeling in Wired and Wireless Networks (PEMWN), pp. 1–6 (2023). https://doi. org/10.23919/PEMWN58813.2023.10304906

18. He, K., Zhang, X., Ren, S., Sun, J.: Deep residual learning for image recognition. arXiv:1512.03385 [cs] (2015)

19. Darcet, T., Oquab, M., Mairal, J., Bojanowski, P.: Vision transformers need registers (2023)

20. Belkin, M., Niyogi, P.: Laplacian eigenmaps for dimensionality reduction and data representation. Neural Comput. **15**(6), 1373–1396 (2003). https://doi.org/10.1162/ 089976603321780317

21. Dwivedi, V.P., Joshi, C.K., Luu, A.T., Laurent, T., Bengio, Y., Bresson, X.: Benchmarking graph neural networks (2022)

22. Ronneberger, O., Fischer, P., Brox, T.: U-net: convolutional networks for biomedical image segmentation. In: Navab, N., Hornegger, J., Wells, W.M., Frangi, A.F. (eds.) MICCAI 2015. LNCS, vol. 9351, pp. 234–241. Springer, Cham (2015). https://doi.org/10.1007/978-3-319-24574-4_28

23. Perez, E., Strub, F., de Vries, H., Dumoulin, V., Courville, A.: FiLM: visual reasoning with a general conditioning layer (2017)

24. Brockschmidt, M.: GNN-FiLM: graph neural networks with feature-wise linear modulation (2020)

25. Ha, D., Dai, A., Le, Q.V.: Hypernetworks (2016)

26. Fioretto, F., Van Hentenryck, P., Mak, T.W.K., Tran, C., Baldo, F., Lombardi, M.: Lagrangian duality for constrained deep learning (2020)

27. Ho, J., Jain, A., Abbeel, P.: Denoising diffusion probabilistic models (2020)

# Enhancing Social Media Profile Authenticity Detection: A Bio-Inspired Algorithm Approach

Nadir Mahammed[1]([✉]) [iD], Badia Klouche[1], Imène Saidi[1] [iD], Miloud Khaldi[1],
and Mahmoud Fahsi[2] [iD]

[1] LabRI-SBA Laboratory, Ecole Superieure en Informatique Sidi Bel Abbes,
P.O 73, El Wiam, 22016 Sidi Bel Abbés, Algeria
{n.mahammed,b.klouche,i.saidi,m.khaldi}@esi-sba.dz
[2] EEDIS Laboratory, Djillali Liabes University, P.O 89, 22000 Sidi Bel Abbès,
Algeria
mahmoud.fahsi@univ-sba.dz

**Abstract.** In the contemporary digital landscape, the pervasive and far-reaching impact of online social networks is indisputable. Prominent platforms such as Instagram, Facebook, and Twitter frequently grapple with the persistent challenge of distinguishing between registered profiles and genuinely engaged users, resulting in a noticeable surge in the prevalence of counterfeit or dormant accounts. This situation underscores the compelling necessity to accurately differentiate between authentic and misleading user profiles. The primary objective of this investigation is to introduce an innovative approach to profile validation. This unique method astutely leverages state-of-the-art bio-inspired algorithms while circumventing traditional machine learning techniques. The empirical results are notably convincing, consistently achieving a high level of accuracy in classification tests conducted on the provided datasets.

**Keywords:** Online social network · Fake profile detection · Bio-inspired algorithm · Machine learning · Instagram · Facebook · Twitter · Simulation

## 1 Introduction

In the ever-evolving landscape of online social networks, as exemplified by the behemoths Instagram, Facebook, and Twitter, a remarkable surge in user engagement has occurred over recent years. This rapid growth, however, has been accompanied by a troubling escalation in the presence of fake accounts and online impersonation. This issue is not only on the rise but has also gained significant scholarly attention, as evident in [5] report on detecting fake profiles. The essence of these fake profiles lies in their representation of fictitious personas or entities that expertly mimic real users, raising pertinent concerns within the online social network ecosystem.

E. Renault et al. (Eds.): MLN 2023, LNCS 14525, pp. 32–49, 2024.
https://doi.org/10.1007/978-3-031-59933-0_3

One of the fundamental challenges in this domain is the absence of robust authentication mechanisms on many social networking platforms. These mechanisms are instrumental in effectively distinguishing between genuine user accounts and fraudulent counterparts. As underscored by [17] in their 2022 survey, the deficiencies in these mechanisms exacerbate the proliferation of fake accounts, thus prompting a dire need for an innovative and effective solution. Such a solution is essential to identify and mitigate the presence of counterfeit accounts, ultimately ensuring the creation of a secure and trustworthy environment for the multitude of users frequenting social networking sites.

In response to this pressing need, the authors of this study have embarked on a transformative journey. They have boldly departed from the well-trodden path of Machine Learning (ML) methods and ventured into the promising realm of metaheuristics. Within this realm, they've harnessed the capabilities of the Fire Hawk Optimizer (FHO), a contemporary bio-inspired algorithm, to confront the multifaceted challenge of fake profile detection. This unorthodox approach marks a significant departure from conventional methodologies and stands as a beacon of innovation, promising to revolutionize the domain of online social network analysis.

The ensuing sections of this comprehensive study delve into the foundational tenets and practical implications of this pioneering approach. By shedding light on its diverse facets, the study aims to emphasize the transformative potential of FHO in the context of enhancing the security and authenticity of online social networks on a global scale. It is, therefore, not just a theoretical endeavor but a promising catalyst for substantial change in the landscape of social network analysis and the wider digital sphere.

## 2   Related Work

The forthcoming section delves into an extensive exploration of contemporary advancements in fake profile detection within the realm of online social media. This exploration extends beyond the confines of conventional machine learning techniques, shedding light on a multifaceted landscape of innovative approaches and methods.

In [13], the growing issue of fake profiles on popular online social networks like Facebook and Twitter is addressed. The paper presents a novel approach to detect fake profiles by combining a machine learning algorithm with a bio-inspired algorithm. The hybrid approach involves two stages, with the first stage utilizing the Satin Bowerbird Optimization algorithm (SBO) to find the best initial centroid for the k-means clustering algorithm in the second stage. This approach demonstrates superior performance compared to established machine learning algorithms in the context of fake profile detection, emphasizing the importance of ensuring the authenticity and security of online social network interactions.

The authors of [8] address the issue of fake online reviews and ratings that mislead customers in their purchasing decisions. They introduce two novel deep-learning hybrid techniques, CNN-LSTM for detecting fake online reviews and

LSTM-RNN for detecting fake ratings. These models outperform existing methods and achieve high prediction accuracy with a particular focus on Amazon datasets.

The main idea is that [20] is addressing privacy concerns, pitching to brands and marketers, expanding the prototype, enabling bulk image processing, integrating with social media APIs, improving the model's recall for logo detection, increasing generalizability, exploring localization of logos, combining textual analytics, automating model selection and hyper-parameter tuning, and comparing performance with existing logo detection systems.

The research of [18] introduces the Multi-Relational Graph-Based Twitter Account Detection Benchmark (MGTAB) to enhance social media user stance and bot detection methods. MGTAB overcomes issues of low annotation quality and incomplete user relationships in existing benchmarks by providing a comprehensive dataset of 1.55 million users and 130 million tweets.

[16] develops a method for identifying and verifying duplicate profiles in online social networks. They use attribute and network-based similarity measures, implement their model with MapReduce to reduce computational complexity, and create a dataset for testing. The study employs parallel k-means clustering and parallel SVM classification techniques to effectively identify suspicious profiles within clusters containing genuine ones.

The [4] work introduces a novel technique called GWODS for detecting attacker shilling profiles in recommender systems. GWODS combines the K-means clustering algorithm with the Grey Wolf Optimizer (GWO) to identify suspicious profiles. It shows promising results in experiments on MovieLens datasets and can be used as a preprocessing step to prevent biased recommendations in recommender systems.

In [12] article, the main idea is that social media bot detection, particularly on Twitter, faces limitations in data collection methods. While machine learning-based algorithms exhibit near-perfect performance on existing datasets, the study reveals that accuracy is often due to dataset-specific factors rather than inherent differences between humans and bots. Additionally, the use of decision trees is preferred due to their interpretability over random forest classifiers.

Table 1 provides a comprehensive overview of the state of research in fake profile detection, emphasizing the need for further investigations that integrate various techniques, improve generalization, and address the dynamic nature of online threats in OSNs. Fake profile detection is a critical aspect of maintaining the integrity and security of online platforms, and these studies play a crucial role in advancing the field.

- Research diversity: The table highlights a diverse range of research efforts directed towards fake profile detection. These studies demonstrate the increasing recognition of the severity of fake profiles in OSNs and the urgency to tackle this issue. This diversity is encouraging as it suggests that multiple avenues are being explored to address the problem.
- OSN-specific approaches: Several studies target specific OSNs, such as Facebook, Instagram, and Twitter. OSN-specific approaches acknowledge the

**Table 1.** Related work summary

| Reference | OSN | ML | Metaheuristic | Other | Dataset | Results (acc) |
|---|---|---|---|---|---|---|
| [13] | Facebook | ID3, SVM, NB, RF, KNN and K-means | SBO | – | 1244 | 0.98 |
| [8] | – | – | – | CNN, LSTM, RNN | 20000 | 0.93 |
| [20] | Instagram | – | – | CDS | 10000 | 0.90 |
| [18] | Twitter | RF, DT, SVM | – | Adaboost | 130 millions | 0.97 |
| [16] | Facebook | SVM, K-means | – | MapReduce | 1000 | 0.98 |
| [4] | - | k-means | GWO | – | 6000 | 0.99 |
| [12] | Twitter | DT, RF | – | – | – | 0.91 |

unique characteristics and challenges posed by each platform. This raises the question of whether a universal model can effectively detect fake profiles across various OSNs or if tailored solutions are necessary.

- Machine learning and metaheuristics: The employed techniques include traditional machine learning algorithms like Decision Trees (DT), Random Forest (RF), Support Vector Machine (SVM), and K-means, as well as bio-inspired metaheuristics like Satin Bowerbird Optimization (SBO) and Grey Wolf Optimizer (GWO). This blend of methodologies indicates the exploration of both data-driven and heuristic-driven approaches. Evaluating their relative efficacy and understanding when to use each type of algorithm is a potential research avenue.

- Incorporation of deep learning: Some studies incorporate deep learning methods, such as Convolutional Neural Networks (CNN), Long Short-Term Memory (LSTM), and Recurrent Neural Networks (RNN). These techniques excel in capturing complex patterns in data. Their integration underscores the need for more advanced methods to combat sophisticated fake profiles, as some may employ deep learning in their creation.

- Dataset size and quality: Dataset size is pivotal in fake profile detection, with some studies using datasets containing millions of instances. Larger datasets can provide more robust training but also demand greater computational resources. Additionally, dataset quality is vital, as low-quality or noisy data can affect model performance. Research into dataset collection and curation techniques is essential.

- Accuracy achievements: Notably, some studies have achieved very high accuracy levels (e.g., 0.98 and 0.99). While high accuracy is a promising sign, it's vital to examine the generalization capabilities of these models. Achieving such high accuracy on one dataset does not necessarily guarantee success on new, unseen data.

- Challenges and future directions: Challenges in fake profile detection include the constant evolution of fake profile creation techniques and the need for real-time or near-real-time detection. Future research should address these challenges and explore methods for dynamic model adaptation.
- Integration and model ensemble: Combining the strengths of different models from various studies or creating ensemble models can potentially improve detection accuracy. Research in this direction could lead to more robust solutions.
- Explainability and interpretability: As fake profile detection systems are deployed, there's a growing need for interpretability and explainability in model decisions, especially in legal and ethical contexts.
- Scalability: Ensuring that fake profile detection methods are scalable to handle the ever-increasing volume of data on OSNs is a significant concern. Research should focus on the efficiency of algorithms in large-scale scenarios.

## 3   Material and Methodology

### 3.1   Datasets

The first dataset was sourced from the online social network Instagram[1]. It consists of 696 instances and encompasses 11 attributes, as specified in [9]: The presence of a profile picture, the ratio of numbers to username length, the count of words in the full name, the ratio of numbers to full name length, the match between the name and username, the length of the description, the presence of an external URL, the privacy status, the number of posts, the number of followers, the number of accounts followed, and the classification as a fake account.

The second dataset employed in this article was constructed from the online social network Facebook. This dataset comprises 1,244 instances and encompasses 15 features, including 'id', 'name-id', 'link', 'profile picture', 'number of likes', 'number of groups joined', 'number of friends', 'education status', 'work', 'living place', 'relationship', 'check-in', 'number of posts', 'number of tags', and 'profile intro', as detailed in [2]. The distribution of its instances is delineated in the accompanying Table 2.

The third dataset was generated through the Twitter API. The Twitter API consists of four main objects: Tweets, Users, Entities and Places. These objects have various attributes [10]. Tweets are the basic building blocks, offering information such as creation time, likes, and retweets, although these attributes are unavailable for protected accounts. Users represent individuals or entities and provide data like liked tweets, followers, and accounts followed. These attributes are accessible for both protected and public accounts. Entities offer additional contextual data within tweets, including hashtags, media, and URLs. Lastly, Places are named locations associated with geographic coordinates. The dataset contains 16 attributes, with a focus on those from Users due to their widespread availability [10].

---

[1] https://urlz.fr/o7Zj.

**Table 2.** Datasets distribution

| Social network | Distribution | Real profile | Fake profile |
|---|---|---|---|
| Instagram | Record | 348 | 348 |
| | Ratio (%) | 50 | 50 |
| Facebook | Record | 1043 | 201 |
| | Ratio (%) | 83.84 | 16.16 |
| Twitter | Record | 499 | 501 |
| | Ratio (%) | 49.9 | 50.01 |

## 3.2  Preprocessing

Data preprocessing is the process of converting raw data into a format that can be readily understood by machine learning algorithms. As detailed in [11], the data preparation procedures for the different datasets employed in this research are succinctly outlined below:

1. Data Scrutiny: Eliminate duplications and rectify errors.
   (a) Eliminate duplications, superfluous data points, inaccuracies, and redundant columns (such as 'id' and 'id-name').
   (b) Omit irrelevant data points, inaccuracies, and redundant columns (such as 'id' and 'id-name').
2. Address disparities, anomalies, and missing data.
3. Standardize and adapt the data through scaling.
4. Prune interrelated variables and streamline the dataset.

## 3.3  Machine Learning Algorithms

**Induction of Decision Tree.** When it comes to decision tree induction, it's worth noting that ID3 is a supervised learning algorithm. This algorithm constructs a tree based on information derived from the training instances and employs it for classifying the test data [7].

**K-Means Algorithm.** A staple in unsupervised learning for pattern recognition and machine learning, is renowned for its simplicity and widespread use among iterative and hill-climbing clustering algorithms [19].

**Hierarchical Clustering Analysis.** HC groups similar objects into clusters. It starts with each object as a separate cluster and then iteratively merges the closest clusters until they all form a single, hierarchical structure. This method is valuable for revealing data patterns and relationships [14].

**Nearest Neighbor Classification.** Often referred to as K-nearest neighbors (KNN), is grounded in the concept that the nearest patterns to a target pattern, for which we are seeking a label, offer valuable label information [15].

**Naive Bayes Classifier.** Commonly referred to as NB, is a supervised learning algorithm rooted in Bayes' theorem. It operates on the simplifying assumption that attribute values are conditionally independent when considering the target value [1].

**Random Forest Machine.** Random forests (RF) represent an amalgamation of tree predictors, where each tree relies on the values of a random vector, independently sampled with a uniform distribution shared across all trees within the forest [6].

**Support Vector Machine.** It is customary to recognize the Support Vector Machine (SVM) as a potent tool for classifier construction. SVM is purposefully designed to establish a robust decision boundary between two classes, thereby facilitating the accurate prediction of labels from one or more feature vectors [21].

### 3.4 Proposed Algorithm

**Inspiration.** Native Australians have a longstanding tradition of using fire for ecosystem management. While fires are intentionally started or ignited by lightning, they can spread due to various factors, impacting the landscape and wildlife. Surprisingly, Fire Hawks, including whistling kites, black kites, and brown falcons, have recently been recognized as contributors to fire spread. These birds deliberately carry burning sticks to set fires as a predatory strategy. They use these fires to startle and capture prey like rodents, snakes, and other animals, making their hunting more efficient.

**Motivation to Choose.** This bio-inspired approach, honed through eons of evolution, equips FHO for intricate optimization. FHO excels in rapid convergence, outperforming alternatives. Its robustness allows it to handle noisy, uncertain data effectively. Moreover, the Fire Hawk Optimizer (FHO) enhances diversity in solution exploration.

FHO's remarkable convergence speed proves valuable in time-sensitive or resource-constrained applications. It reaches optimal solutions swiftly. The process iterates until predefined criteria are met. FHO's computational efficiency is evident as it converges to the global best with fewer evaluations [3].

**Operation.** The FHO algorithm, inspired by the foraging behavior of fire hawks, operates through the following steps:

1. Initial Positioning: At the start, solution candidates (X) are defined, representing the positions of fire hawks and prey in the search space. Random initialization places these vectors within the search space, taking into account various parameters.

2. Fire Hawks and Prey: The algorithm categorizes solution candidates into Fire Hawks and prey based on their objective function values. Selected Fire Hawks aim to spread fires around the prey, with the global best solution serving as the primary fire source.
3. Determining Territories: The algorithm calculates the total distance between Fire Hawks and prey to identify the nearest prey to each bird. This step determines the effective territory of the Fire Hawks for hunting. The bird with the best objective function value selects the nearest prey to its territory, while others choose their next nearest prey.
4. Spreading Fires: Fire Hawks collect burning sticks from the main fire and drop them in their territories, causing the prey to flee. Some Fire Hawks may use burning sticks from other territories, contributing to position updates in the search loop.
5. Prey Movements: The prey's movements within Fire Hawks' territories are considered. The algorithm simulates various prey actions, such as hiding, running, or approaching Fire Hawks, impacting position updates.
6. Safe Places: Prey may move toward safe places outside Fire Hawk territories. These movements are also included in the position update process.
7. Territory Definition: Fire Hawk territories are represented as circular areas, with the precise territory determined by prey numbers and distances from each Fire Hawk.
8. Boundary Violation and Termination: The algorithm considers boundary control for violating decision variables and employs a termination criterion, such as a predefined number of objective function evaluations or iterations, to conclude the process.

The Fig. 1 provides pseudocode which offers a concise overview of the FHO algorithm's operation.

**Transition from Natural to Artificial.** This section is devoted to examining the shift from the Fire Hawk's innate behaviors in the wild to its adapted behaviors in an artificial environment, as detailed in the Table 4.

Table 4 delves into a captivating comparison between the natural and artificial, spotlighting the FHO algorithm's mission of distinguishing genuine from fraudulent profiles in online social networks. It intriguingly parallels the hunting behavior of fire hawks with user suitability assessment.

By mentioning distance calculations, it hints at the algorithm's quest for the optimal solution, equating to precise user classifications in social networks. This table is a gateway to understanding how nature's wisdom inspires advanced algorithms that address real-world challenges.

It embodies the fusion of the natural and artificial realms, demonstrating how algorithmic innovation stems from nature's timeless principles, resolving complex issues in online social networks. Ultimately, it invites exploration of the limitless possibilities born from the fusion of nature and algorithms (Table 3).

```
Fire Hawk Optimizer
        Determine initial positions of solution candidates (Xᵢ) in the search space with N candidates
        Evaluate fitness values for initial solution candidates
        Determine the Global Best (GB) solution as the main fire
        While Iteration < Maximum number of iterations
              Generate n as a random integer number for determining the number of Fire Hawks
              Determine Fire Hawks (FE) and Preys (PR) in the search space
              Calculate the total distance between the Fire Hawks and the preys
              Determine the territory of the Fire Hawks by dispersing the preys
              For l=1: n
                    Determine the new position of the Fire Hawks
                    For q=1:r
                          Calculate the safe place under lᵗʰ Fire Hawk territory
                          Determine the new position of the preys
                          Calculate the safe place outside the lᵗʰ Fire Hawk territory
                          Determine the new position of the preys
                    End
              End
              Evaluate fitness values for the newly created Fire Hawks and preys
              Determine the Global Best (GB) solution as the main fire
        End While return GB
```

Fig. 1. FHO Pseudocode.

Table 3. Transition from natural to artificial of FHO

| Natural | Artificial |
|---------|------------|
| Natural Artificial Fire hawks hunting for prey in the wild | Each user is classified into the most Suitable class (Real or Fake) |
| Fire hawks finding food by following the smoke signals from wildfires | Two classes (Real or Fake) |
| Environment | Online Social Networks (Facebook, Twitter, Instagram) |
| Fire hawk | Online Social Networks User |
| Group of fire hawks | Online Social Networks Users |
| Best individual in the group of fire hassle that found the prey | The best solution in the population of solutions that meets the objective function by the FHO (Real or Fake) |
| The distance between the flee hawk and its prey | $D_k^1 = \sqrt{(x_2 - x_1) + (y_2 - y_1)}$ The distance between the solution and the optimal |

**Fitness Function.** The FHO rigorously employs a fitness function, as depicted in Fig. 2, to meticulously gauge the performance of solution candidates. This fitness function pivots around the precision of a gradient boosting classifier meticulously applied to a thoughtfully selected subset of features sourced from a dataset.

To elaborate on the computation of the fitness value, the function takes a solution candidate into its fold, representing a distinct subset of features. This subset undergoes scrupulous evaluation via a gradient boosting classifier, armed with precisely 100 estimators and a deterministic random state fixed at 42. Notably, this classifier undertakes the dual responsibility of feature selection and classification.

The inner workings of the fitness function encompass the formulation of a feature selector. This selector, entailing sophisticated intricacies, leverages the classifier itself to discern and pinpoint the paramount features based on the classifier's predictive capabilities. This discernment is crucial in optimizing the classification process.

Of particular significance is the selector's subsequent fitting to both the input dataset and the target variable. This preparatory phase is pivotal for the forthcoming accuracy evaluation.

What distinguishes this fitness function is its intrinsic capacity to bring about a transformation of the input dataset. This transformation is rendered by carefully cherry-picking the most pivotal features from the original dataset. The result is a transformed dataset, which bears the promise of enhanced accuracy. This transformed dataset now becomes the testing ground for the classifier. It serves as the substrate for the classifier's extensive training process, conducted in close tandem with the target variable.

As the final step in this intricate dance of precision, the fitness function introduces the crucial concept of the accuracy score. It orchestrates a meticulous comparison between the true labels and the predicted labels that emerge from the classifier's outputs on the transformed dataset. The resultant accuracy score stands as a testament to the chosen subset of features' ability to effectively forecast the target variable.

```
Fitness Function
    1: def fitness function(solution):
    2: classifier = GradientBoostingClassifier(n estimators=100, random state=42)
    3: selector = SelectFromModel(classifier)
    4: selector.fit(X, y)
    5: X selected = selector.transform(X)
    6: classifier.fit(X selected, y) ▷Return the accuracy score on the training set
    7: return accuracy score(y, classifier.predict(X selected))
```

Fig. 2. Fitness Function.

Figure 3 demonstrates the pivotal role of the fitness function in the FHO. In the third stage of the code, the fitness values for each solution candidate in the population are meticulously computed by invoking the fitness function. This function is systematically applied to every row (axis=1) within the population array, yielding an array replete with fitness values, which are more specifically accuracy scores. These accuracy scores bear significance as they provide a quantitative assessment of each solution candidate's performance accuracy.

In essence, the fitness function operates as the core evaluator, discerning and ranking solution candidates based on their individual performance. In the broader context, these fitness scores wield substantial influence in steering the FHO's pursuit of the optimal solution, with the overarching goal of optimizing performance accuracy.

---

**Fitness Function within FHO**

\# Determine the search space and initialize solution.

1: search space = (0, X.shape[1]) candidates

2: population = np.random.rand(self.population size, X.shape[1])

\# Evaluate fitness values for initial solution candidates.

3: fitness values = np.apply along axis(fitness function, 1, population)

---

**Fig. 3.** Fitness Function with FHO.

**FHO Metrics.** The FHO algorithm undergoes a comparative analysis against a spectrum of established Machine Learning algorithms, encompassing ID3, SVM, NB, RF, HC, KNN with diverse K values, and K-means. This exhaustive evaluation consists of 100 iterations for each dataset, ensuring robustness and careful examination. Notably, the FHO configuration parameters are as follows: the initial population size is set at 50, and the maximum number of iterations is capped at 100 as summarized in Table 4.

**Table 4.** FHO metrics

| Parameter | Value |
|---|---|
| Population size | 50 |
| Iteration | 100 |
| Iteration by dataset | 100 |

## 4    Results and Discussion

Throughout the experimental phase, a 2014 MSI GT70 gaming laptop was employed, featuring an Intel Core i7-4800MQ CPU, a Nvidia GeForce GTX 770M GPU, and 32 GB of RAM.

## 4.1 Evaluation Criteria

The detection of fake accounts can be evaluated using various performance metrics, such as Accuracy, F-score, Recall, precision, and entropy. These metrics provide insights into the model's performance and its ability to classify profiles correctly.

In addition, the Confusion Matrix is used as a visual representation of fake account detection, offering a comprehensive view of the model's performance across different classes as shown in Table 5.

**Table 5.** Results classes

| Class | Meaning |
|---|---|
| True Positives (TP) | Instances where the model accurately identifies positive cases within the dataset |
| False Positives (FP) | Cases in which the model incorrectly categorizes negative instances as positives |
| True Negatives (TN) | Cases in which the model correctly recognizes negative instances |
| False Negatives (FN) | Instances in which the model erroneously categorizes positive cases as negative |

– Accuracy: This metric measures the overall accuracy of the model in correctly classifying profiles.

$$Accuracy = \frac{TP + TN}{TP + TN + FP + FN} \tag{1}$$

– Precision: Calculates the model's accuracy in classifying values correctly by comparing the number of accurately classified profiles to the total classified data points for a given class label.

$$Precision = \frac{TP}{TP + FP} \tag{2}$$

– Recall: This metric assesses the model's ability to correctly predict positive values, indicating how often it correctly identifies true positives.

$$Recall = \frac{TP}{TP + FN} \tag{3}$$

– F1-score: Which is the harmonic mean of precision and recall, balances the trade-off between these two metrics.

$$F1 - score = \frac{2 * TP}{2 * TP + FP + FN} \tag{4}$$

- Entropy: This metric quantifies the randomness or disorder in a system, providing valuable information about the data's structure and organization.

$$Entropy = log2(Precision) * (-Precision) \qquad (5)$$

## 4.2   Results

Tables 6, 7 and 8 offer a comprehensive view of the results obtained using datasets sourced from Instagram, Facebook, and Twitter. It's important to highlight that a rigorous cross-validation technique was employed to gauge the effectiveness of the chosen machine learning methods when compared to the proposed bio-inspired algorithm. Through this approach, the dataset was thoughtfully divided into two segments: 70% for training and 30% for testing.

**Table 6.** Instagram results.

|  | Accuracy | Precision | Recall | F1-score | Entrorpy |
|---|---|---|---|---|---|
| **FHO** | **0.96** | **0.96** | **0.96** | **0.96** | **0.02** |
| ID3 | 0.86 | 0.86 | 0.85 | 0.85 | 0.06 |
| K-means | 0.40 | 0.30 | 0.36 | 0.30 | 0.16 |
| HC | 0.64 | 0.73 | 0.66 | 0.62 | 0.10 |
| KNN (k=5) | 0.90 | 0.90 | 0.90 | 0.92 | 0.04 |
| NB | 0.93 | 0.78 | 0.66 | 0.60 | 0.08 |
| RF | 0.94 | 0.94 | 0.94 | 0.94 | 0.03 |
| SVM | 0.91 | 0.91 | 0.91 | 0.91 | 0.04 |

Table 6 presents a comprehensive overview of the outcomes derived from the analysis of Instagram datasets using a diverse range of classification algorithms. Table 6 meticulously records the performance metrics, such as Accuracy, Precision, Recall, F1-score, and Entropy, for each of the algorithms deployed in the study. The prominently featured FHO algorithm notably achieved an exceptional accuracy score of 0.96, underscoring its proficiency in the task of fake profile detection. Conversely, the outcomes for alternative algorithms, encompassing ID3, K-means, HC , KNN (with k=5), NB, RF, and SVM are also presented. These results provide a nuanced perspective on the varying performances of these techniques when applied to the domain of Instagram profile verification.

In Table 7, the results for Facebook datasets are presented, highlighting the performance of various machine learning algorithms, with a specific focus on the FHO. The latter stands out with exceptional accuracy, precision, recall, F1-score, and entropy values, all at 0.99, surpassing the other algorithms. These results demonstrate FHO's remarkable ability to accurately classify Facebook profiles as genuine or fake. Its exceptionally low entropy of 0.00 indicates minimal randomness and structural inconsistencies within identified fraudulent profiles, emphasizing FHO's robustness in detecting subtle nuances that often elude

**Table 7.** Facebook results.

|  | Accuracy | Precision | Recall | F1-score | Entrorpy |
|---|---|---|---|---|---|
| **FHO** | **0.99** | **0.99** | **0.99** | **0.99** | **0.00** |
| ID3 | 0.96 | 0.97 | 0.97 | 0.97 | 0.01 |
| K-means | 0.60 | 0.60 | 0.60 | 0.60 | 0.13 |
| HC | 0.60 | 0.60 | 0.60 | 0.60 | 0.13 |
| KNN (k=3) | 0.83 | 0.63 | 0.60 | 0.61 | 0.12 |
| NB | 0.96 | 0.97 | 0.96 | 0.97 | 0.02 |
| RF | 0.96 | 0.97 | 0.96 | 0.97 | 0.02 |
| SVM | 0.96 | 0.97 | 0.96 | 0.96 | 0.02 |

other algorithms. This comprehensive assessment underscores FHO's significant potential in the realm of fake profile detection on Facebook, highlighting its profound contributions to enhancing platform security and user trust.

**Table 8.** Twitter results.

|  | Accuracy | Precision | Recall | F1-score | Entrorpy |
|---|---|---|---|---|---|
| **FHO** | **0.96** | **0.96** | **0.96** | **0.96** | **0.01** |
| ID3 | 0.88 | 0.88 | 0.88 | 0.86 | 0.04 |
| K-means | 0.54 | 0.63 | 0.50 | 0.35 | 0.12 |
| HC | 0.53 | 0.27 | 0.5 | 0.35 | 0.15 |
| KNN (k=5) | 0.82 | 0.82 | 0.82 | 0.82 | 0.07 |
| NB | 0.91 | 0.91 | 0.91 | 0.91 | 0.03 |
| RF | 0.92 | 0.92 | 0.92 | 0.92 | 0.03 |
| SVM | 0.79 | 0.79 | 0.79 | 0.79 | 0.8 |

Table 8 provides an in-depth evaluation of the Twitter dataset, emphasizing the performance of various machine learning algorithms, with a specific focus on the FHO. Notably, FHO achieves a high level of accuracy, precision, recall, and F1-score, all standing at 0.96, demonstrating its proficiency in effectively distinguishing between genuine and fraudulent Twitter profiles. FHO also maintains a commendably low entropy of 0.01, indicating the minimal degree of unpredictability or randomness within the profiles it identifies as fake. In contrast to FHO's strong performance, other algorithms display comparatively lower scores in various metrics, highlighting FHO's distinct advantage in detecting counterfeit Twitter accounts. These results underscore FHO's significant potential for enhancing Twitter's platform security and user confidence in the authenticity of profiles, thus contributing significantly to mitigating the prevalence of fake profiles on the platform.

## 4.3 Discussion

Regarding to The FHO, the outcomes portrayed in Tables 6, 7, and 8 draws its computational prowess from a rich tapestry of biological inspiration, firmly establishing itself within the domain of metaheuristic algorithms. Rooted in the primal strategies of fire hawks, adept predators in the wild, FHO seamlessly blends these natural influences with computational processes. In this digital wilderness, FHO mirrors the precision of foraging dynamics, expertly navigating the solution space akin to its avian counterparts who hone in on bountiful areas where prey thrives. FHO's metaheuristic essence shines through as it proudly claims its membership in the coveted metaheuristics family of optimization algorithms, celebrated for their adaptability and efficiency. This quality is exemplified in FHO's pursuit of diverse solution candidates, making it an extraordinary asset in tackling intricate challenges like fake profile detection. Moreover, FHO's flexibility in handling an array of data types, both continuous and discontinuous, strengthens its role as a versatile tool for profile validation. A notable distinction lies in FHO's independence from conventional training or data-splitting processes, championing computational efficiency and immediate problem-solving. Collectively, FHO's bio-inspired roots, metaheuristic character, adaptability, and direct application capability underscore its exceptional performance in the intricate realm of online social network analysis.

The results showcased in Tables 6, 7, and 8 accentuate the remarkable proficiency of the Fire Hawk Optimizer (FHO) when dealing with the datasets from Instagram, Facebook, and Twitter, specifically in the realm of fake profile detection. Across pivotal performance metrics, FHO consistently outperforms alternative machine learning algorithms.

FHO's standout attribute is its remarkable ability to rapidly converge towards predefined tolerance for the global best solution. This swift convergence, coupled with its resource-efficiency, assumes particular significance in the context of social networks where timely profile verification is crucial, and computational resources often come at a premium.

What sets FHO apart is its innate knack for handling the unpredictability and noise inherent in real-world data, showcasing its robustness and adaptability in navigating the often erratic nature of user-generated profile information.

FHO's penchant for diversifying the search process, inspired by natural systems, is another remarkable trait. By concurrently exploring multiple potential solutions, it enhances the likelihood of uncovering innovative answers, a pivotal asset when contending with the ever-evolving strategies employed by creators of fake profiles.

The results speak to FHO's exceptional computational efficiency. It consistently converges to the global best solution within a significantly reduced timeframe, allowing it to swiftly identify optimal or near-optimal solutions. This efficiency proves highly pertinent in situations where time sensitivity and conservation of computational resources are paramount.

An additional noteworthy aspect is FHO's ability to converge toward the global best solution in mathematical test functions while requiring fewer objective function evaluations. This underscores its computational efficiency, highlighting its practical applicability across a spectrum of problem-solving scenarios.

## 5 Conclusion

Amid the landscape of online social media platforms, the issue of fake profiles has emerged as a pressing concern, particularly on major platforms like Instagram, Facebook, and Twitter. The phenomenon of a growing gap between registered profiles and genuinely active users has raised red flags, indicating an alarming increase in counterfeit or inactive accounts. This situation not only jeopardizes the credibility of these platforms but also exposes them to security and privacy risks. Over the years, the academic literature has primarily explored the application of machine learning techniques to address this challenge. These techniques have sought to discern real from fraudulent profiles by analyzing various attributes and user behavior patterns. However, these traditional methods exhibit limitations, necessitating a quest for more robust and efficient solutions.

A transformative shift in the quest to combat fake profiles has taken shape, focusing on the potential of metaheuristic algorithms, with specific emphasis on bio-inspired algorithms. This shift acknowledges the constraints of conventional machine learning approaches in effectively handling the complex realities of online social network data. Bio-inspired algorithms, represented by the Fire Hawk Optimizer, have emerged as promising contenders in the realm of fake profile detection. FHO derives its computational prowess from its inherent bio-inspired nature, which draws inspiration from the foraging behavior of fire hawks.

The metaheuristic dimension of FHO amplifies its significance. As a member of the metaheuristics family, FHO belongs to a class of optimization algorithms lauded for their adaptability and efficiency. FHO sets itself apart through its pursuit of diverse solution candidates, making it adept at addressing complex, multifaceted challenges, especially in the context of fake profile detection.

The proficiency of FHO is vividly demonstrated through the performance results associated with Instagram, Facebook, and Twitter datasets. Notably, FHO excels in promptly and efficiently converging toward the global best solution, a trait of paramount importance in scenarios where timely profile validation and limited computational resources are critical. Moreover, the algorithm resilience in handling unpredictable data and its ability to diversify the search process are valuable assets when confronting the ever-evolving tactics employed by creators of fake profiles. Additionally, its computational efficiency, characterized by a lower number of objective function evaluations while consistently converging to the global best solution, positions it as an exemplar of computational prowess.

Regarding the importance of the subject, several future perspectives can be considered. Refining and advanced FHO's capabilities by adapting it to large datasets with heterogeneous data. The integration of FHO with other advanced techniques is a compelling avenue. Future studies could explore hybrid approaches that leverage the strengths of FHO alongside complementary methods for even more robust profile validation.

# References

1. Abdulkareem, N.M., Abdulazeez, A.M., Zeebaree, D.Q., Hasan, D.A.: COVID-19 world vaccination progress using machine learning classification algorithms. Quba-han Acad. J. **1**(2), 100–105 (2021)
2. Albayati, M., Altamimi, A.: MDFP: a machine learning model for detecting fake Facebook profiles using supervised and unsupervised mining techniques. Int. J. Simul.: Syst. Sci. Technol. **20**(1), 1–10 (2019)
3. Azizi, M., Talatahari, S., Gandomi, A.H.: Fire hawk optimizer: a novel metaheuristic algorithm. Artif. Intell. Rev. **56**(1), 287–363 (2023)
4. Bansal, S., Baliyan, N.: Detecting group shilling profiles in recommender systems: a hybrid clustering and grey wolf optimizer technique. In: Singh, D., Garg, V., Deep, K. (eds.) Design and Applications of Nature Inspired Optimization. Women in Engineering and Science, pp. 133–161. Springer, Cham (2023). https://doi.org/10.1007/978-3-031-17929-7_7
5. Bhambulkar, R., Choudhary, S., Pimpalkar, A.: Detecting fake profiles on social networks: a systematic investigation. In: 2023 IEEE International Students' Conference on Electrical, Electronics and Computer Science (SCEECS), pp. 1–6. IEEE (2023)
6. Biau, G., Scornet, E.: A random forest guided tour. Test **25**, 197–227 (2016)
7. Charbuty, B., Abdulazeez, A.: Classification based on decision tree algorithm for machine learning. J. Appl. Sci. Technol. Trends **2**(01), 20–28 (2021)
8. Deshai, N., Rao, B.B., et al.: Deep learning hybrid approaches to detect fake reviews and ratings. J. Sci. Industr. Res. **82**(1), 120–127 (2022)
9. Dey, A., Reddy, H., Dey, M., Sinha, N., Joy, J.: Detection of fake accounts in Instagram using machine learning. Int. J. Comput. Sci. Inf. Technol. **11**(5), 83–90 (2019)
10. Erşahin, B., Aktaş, Ö., Kılınç, D., Akyol, C.: Twitter fake account detection. In: 2017 International Conference on Computer Science and Engineering (UBMK), pp. 388–392. IEEE (2017)
11. García, S., Ramírez-Gallego, S., Luengo, J., Benítez, J.M., Herrera, F.: Big data preprocessing: methods and prospects. Big Data Anal. **1**(1), 1–22 (2016)
12. Hays, C., Schutzman, Z., Raghavan, M., Walk, E., Zimmer, P.: Simplistic collection and labeling practices limit the utility of benchmark datasets for twitter bot detection. In: Proceedings of the ACM Web Conference 2023, pp. 3660–3669 (2023)
13. Mahammed, N., Bennabi, S., Fahsi, M., Klouche, B., Elouali, N., Bouhadra, C.: Fake profiles identification on social networks with bio inspired algorithm. In: 2022 First International Conference on Big Data, IoT, Web Intelligence and Applications (BIWA), pp. 48–52. IEEE (2022)
14. Murtagh, F., Contreras, P.: Algorithms for hierarchical clustering: an overview, II. Wiley Interdisc. Rev.: Data Min. Knowl. Discov. **7**(6), e1219 (2017)
15. Prasetyaningrum, P.T., Pratama, I., Chandra, A.Y.: Similiarity report: implementation of machine learning to determine the best employees using random forest method (2023)

16. Saravanan, A., Venugopal, V.: Detection and verification of cloned profiles in online social networks using MapReduce based clustering and classification. Int. J. Intell. Syst. Appl. Eng. **11**(1), 195–207 (2023)
17. Shamseddine, J., Malli, M., Hazimeh, H.: Survey on fake accounts detection algorithms on online social networks. In: Daimi, K., Al Sadoon, A. (eds.) ICR 2022. AISC, vol. 1431, pp. 375–380. Springer, Cham (2022). https://doi.org/10.1007/978-3-031-14054-9_35
18. Shi, S., et al.: MGTAB: a multi-relational graph-based twitter account detection benchmark. arXiv preprint arXiv:2301.01123 (2023)
19. Sinaga, K.P., Yang, M.S.: Unsupervised k-means clustering algorithm. IEEE Access **8**, 80716–80727 (2020)
20. Tanniru, V., Bhattacharya, T.: Online fake logo detection system (2023)
21. Tanveer, M., Rajani, T., Rastogi, R., Shao, Y.H., Ganaie, M.: Comprehensive review on twin support vector machines. Ann. Oper. Res. 1–46 (2022)

# A Deep Learning Based Automatic Outdoor Home Video Surveillance Approach

Hamid Reza Tohidypour, Tala Bazzaza$^{(\boxtimes)}$, Yixiao Wang, Panos Nasiopoulos, Vincent Sastra, Bowei Ren, Elbert Ng, Sebastian Gonzalez, and Andrew Shieh

Department of Electrical and Computer Engineering, University of British Columbia, Vancouver, Canada
{htohidyp,tbazzaza,yixiaow,panosn}@ece.ubc.ca, {vsastra,bren, ngelbert,gsebass1,shiehand}@student.ubc.ca

**Abstract.** Home video surveillance systems are widely used to track potential criminal activities and serve as a formidable deterrent to potential intruders and wrongdoers. Nonetheless, it's essential to emphasize that these systems very frequently lead to false alarms, which can be quite frustrating for users. The existing automatic video surveillance methods and datasets were initially developed for monitoring public crimes, and they do not generalize well for home surveillance purposes. To this end we proposed a deep learning based automatic home video surveillance approach based on anomaly detection. We captured a representative vide dataset for our outdoor home surveillance system and employed this dataset to train our deep learning model. Our evaluations have shown that our model achieved anomaly detection accuracy of 79.12%, which is impressive considering our dataset size.

**Keywords:** Home video surveillance · Deep Learning · Anomaly detection

## 1 Introduction

In recent years, video surveillance has gained significant attention for its role in outdoor home security, allowing homeowners to monitor their property and enhance safety in their neighborhoods. The common practice in outdoor home video surveillance involves the installation of motion-detecting cameras that trigger alarms upon sensing any movement. These alarms are typically followed by capturing a short video clip of the incident, which is stored locally and may be sent to the homeowner's cell phone if the system is configured to do so. However, these systems often encounter a high rate of false alarms, primarily because outdoor surveillance systems can be triggered by various factors such as animals, moving vegetation, the presence of moving vehicles, or changes in lighting conditions, resulting in unnecessary alerts. This issue can be quite frustrating for users, emphasizing the urgent need for a solution. Thus, there is a significant demand for the development of automatic anomaly detection systems for outdoor home video surveillance that can minimize false alarms.

© The Author(s), under exclusive license to Springer Nature Switzerland AG 2024
É. Renault et al. (Eds.): MLN 2023, LNCS 14525, pp. 50–56, 2024.
https://doi.org/10.1007/978-3-031-59933-0_4

To the best of our knowledge, there is currently no existing work especially proposed for automatic outdoor home surveillance systems. However, several efforts have been made to develop automatic video surveillance systems for detecting and preventing various types of public crimes, including but not limited to abuse, arrest, arson, assault, road accidents, burglary, explosions, fighting, robbery, shootings, theft, shoplifting, and vandalism. These systems often leverage advanced computer vision and machine learning techniques to monitor and analyze video footage for suspicious or criminal activities, helping enhance public safety and security [1–3]. In the work outlined in reference [2], the authors introduced the Robust Temporal Feature Magnitude Learning (RTFM) model for public crime detection. This machine learning model operates under the paradigm of weak supervision and comprises two essential components: a Multi-scale Temporal Network and a Snippet Classifier. Notably, RTFM's weakly supervised nature enables it to be trained using categorized videos alone, without the need for precise labeling of the specific frames depicting abnormal behavior. In a separate endeavor, a multiple instance self-training framework (MIST) was developed to enhance task-specific discriminative representations exclusively using video-level annotations, making it an exceptionally efficient solution for public crime detection [2]. MIST comprises two pivotal components: a multiple instance pseudo-label generator and a self-guided attention boost feature encoder. What sets MIST apart is its distinctive two-stage self-training approach, meticulously tailored for video anomaly detection with the overarching goal of training a specialized encoder dedicated to the task at hand. In [3], the authors proposed self-supervised sparse representation (S3R) framework designed to capture the concept of anomalies at the feature level for public crime detection. This approach leverages the synergistic combination of dictionary learning and supervised learning. Nevertheless, since all of these networks were trained using public crime data, their ability to generalize to the specific scenario of home surveillance may be limited.

To address the above mentioned issues, we propose a deep learning based anomaly detection model specially tailored for automatic outdoor home surveillance. We have captured and acted out various scenarios, encompassing both anomalous and non-anomalous situations in outdoor home surveillance, to create a comprehensive dataset for our training and evaluation phases. The performance evaluation revealed that our method achieved a high level of accuracy when tested on unseen videos.

The remainder of this paper follows the following structure: In Sect. 2, we introduce our methodology, explain our dataset, and detail the network architecture utilized. Section 3 is dedicated to the performance evaluation of our method, along with a discussion of the results. Lastly, in Sect. 4, we offer the concluding remarks for our paper.

## 2  Our Approach

Our objective is to develop an automated home outdoor video surveillance system using deep learning technology. The success of our system heavily relies on the availability of a representative and comprehensive dataset that aligns with our application, as well as employing a deep learning structure that is tailored to our specific needs. The following subsections describe our dataset collection and labelling as well as our deep learning model.

a)                                          b)

**Fig. 1.** Examples of a) normal and b) anomalous frames.

## 2.1 Data Collection and Labeling

To the best of our knowledge, there is currently no dedicated dataset specifically captured for home video surveillance purposes. To this end, we created our own dataset by capturing videos in three different outdoor home settings, where both anomalous and non-anomalous events were acted out. Anomalous scenarios include package thefts, suspicious runs, sneaking around, staying long and armed. Normal scenarios include package/food delivery, animals and kids. We subsequently categorize these videos into two primary groups: anomalous and non-anomalous, followed by further subdivision into sub-categories that specify the type of event depicted in the video (e.g., armed incidents, deliveries, package theft, etc.). For each of the anomalous videos, we labeled the frames in the video in which it is anomalous. This is used by the model as "ground truths" to evaluate their own performance. We collected a total of 116 videos for our dataset, comprising 57 anomalous events and 59 normal events. Videos on average are 15–20 s each. The dataset is split into 82 videos for training (70 videos) and 34 videos for validation. Balanced number of anomalous and non-anomalous videos were used for each set. Figure 1 shows two examples of normal and anomalous frames.

## 2.2 Our Deep Learning Network

We investigated the performance of three state-of-the-art methods proposed for public crime detection, including S3R, MIST, and RTFM to find the best network structure for our application. These methods showed great performance on three public crime dataset, namely UCF-crime [4], ShanghaiTech [5], and XD-Violence [6]. Since these datasets are for public crimes and are not representative for the case of home surveillance, we collected a representative test outdoor home surveillance dataset from YouTube to test the pretrained models of the three above mentioned network, which were originally trained on UCF-crime dataset. This dataset consists of 50 videos, different from the dataset mentioned in the previous subsection. 25 of these videos are normal, while the rest were anomalous videos. Table 1 shows the accuracy performance of the three pre trained models. The S3R model stands out as the most consistently performing model among the three under evaluation. Upon closer examination of its frame-by-frame prediction output, it demonstrated strong capabilities in identifying anomalous events. Nevertheless, we observed a notable issue with the model's sensitivity, leading to a relatively high rate of false positives. Notably, it struggles when attempting to predict non-anomalous events within our standard video footage, such as those featuring animals, children, or delivery personnel. This outcome aligns with our expectations, considering that the non-anomalous events in our test videos, such as armed incidents and home invasions, bear some resemblance to the scenarios the model was trained on (specifically, the UCF-Crime dataset). It's important to highlight that the normal videos in our testing dataset occur in a markedly different setting compared to the more public environments depicted in the UCF-Crime dataset videos. Considering S3R's performance, we decided to use this network for our application and fine-tune its pretrained network using our dataset explained in the previous subsection. Since our YouTube dataset was used for choosing network, we did not use it to fine-tune the S3R's model.

**Table 1.** Accuracy of pretrained models on our YouTube dataset

| Methods | S3R | MIST | RTFM |
|---|---|---|---|
| Accuracy | 48.20% | 45.50% | 37.33% |

Figure 2 shows the structure of S3R. The S3R network integrates dictionary learning with self-supervised techniques to capture the concept of feature-level anomalies. Initially, a feature extractor denoted as E represents each untrimmed video x by extracting snippet-level features F. Subsequently, all normal training videos, represented as $\tilde{X}$, are gathered to construct a task-specific dictionary. Following this, the en-Normal module utilizes both F and the dictionary to reconstruct the feature set $\hat{F}$. Subsequently, the de-Normal module analyzes the distinctions between F and $\hat{F}$ to effectively filter out patterns associated with normal events. Ultimately, the filtered features are well-prepared to discern between normal and anomalous events within both the snippet-level and video-level features.

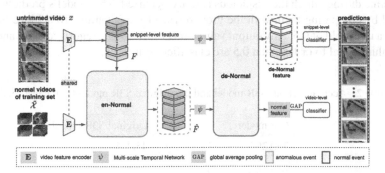

**Fig. 2.** The S3R network structure [3].

To augment the dataset, we incorporated two data augmentation techniques: random horizontal flips and color jitter. Random horizontal flips involve randomly flipping videos horizontally, while color jitter introduces random adjustments to brightness, contrast, and

**Fig. 3.** The output corresponding to an example of suspicious activity. The red region represents the timeline during which the suspicious activity occurred.

saturation in the video. Importantly, as the occurrence of abnormality is not contingent upon the horizontal orientation or the color scheme of the video, these two methods not only expand the dataset but also improve the model's resilience against inconsequential factors, such as video saturation and contrast. We fine-tuned the pretrained model of S3R using our training dataset (explained in the previous section) using a single NVIDIA V100 GPU equipped with 32 GB of memory, which was made accessible through a cutting-edge research computing network [7]. Performance evaluation showed that our method achieved the accuracy of 84.97% for our validation set.

## 3   Evaluation and Discussion

For the testing phase, we recorded our own videos in a home setting different from the ones used for the training and validation. Similar to the training and validation sets both anomalous and non-anomalous events are acted out. We categorized each frame within the video as either anomalous or non-anomalous. An equal number of anomalous and non-anomalous videos were included in each set. We used our test dataset to evaluate the performance of our model and the original S3R model.

Our model outputs a prediction every 16 frames. Figure 3 displays the output corresponding to an example of suspicious activity for our model. The red region indicates the time-lime in which the suspicious activity was performed. The red region represents the timeline during which the suspicious activity occurred. The model's predictions are assessed by comparing them with the ground truth for evaluation. To achieve this, we have set a threshold of 0.5. Any output value exceeding 0.5 is categorized as an anomaly, while values equal to or less than 0.5 are classified as normal.

**Table 2.** Accuracy of our S3R model and the original S3R model for the test sets.

|          | Our model | Original S3R model |
|----------|-----------|--------------------|
| Accuracy | 79.12%    | 45.23%             |

Table 2 shows the accuracy performance of our fine-tuned S3R model and the original S3R model, which was trained for public crime. As can be seen our model achieved the test accuracy of 79.12%, while the original S3R model achieved the test accuracy of 45.23%. Given the relatively small size of our training dataset, the achieved level of accuracy is indeed impressive. Figure 4 showcases multiple instances of our model

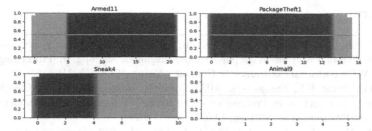

**Fig. 4.** Examples of correct classifications of our model on our test data.

correctly classifying data within our test dataset. These example videos specifically cover four distinct scenarios: the observation of an armed person, package theft, a person sneaking, and the observation of an animal. The red region indicates the time-lime in which the suspicious activity was performed. Figure 5 shows an example of miss classification, in which delivering a package is considered an anomaly.

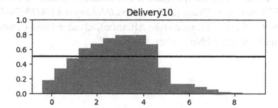

**Fig. 5.** Example of missclassification of our model on our test data.

## 4   Conclusions

In this paper, we proposed a novel deep learning-based automatic surveillance system for outside homes to track potential criminal activities and mitigate the issue of frequent false alarms found in existing home surveillance systems. In this regard, we investigated the performance of the state-of-the-art video surveillance networks, which were proposed for public crime detection, for the home surveillance application. Our preliminary results showed that self-supervised sparse representation (S3R) network model trained on a public crime structure showed promising results for the outdoor home surveillance. However, it achieved the accuracy of less than 50%, which is too low for binary classification. To fine-tune the S3R model for our application, we captured a representative dataset consists of anomalous and non-anomalous events for outdoor home surveillance. We fine-tuned the S3R model using our dataset and augmentation approaches that we introduced for our application. Our evaluations have shown that our model achieved the anomaly detection accuracy of 79.12%. Future work involves increasing the training dataset size by capturing more videos from different home setups.

**Acknowledgments.** This work was supported in part by the Natural Sciences and Engineering Research Council of Canada (NSERC – PG 11R12450), and TELUS (PG 11R10321).

# References

1. Tian, Y., Pang, G., Chen, Y., Singh, R., Verjans, J.W., Carneiro, G.: Weakly-supervised video anomaly detection with robust temporal feature magnitude learning. In: Proceedings of the IEEE/CVF International Conference on Computer Vision (ICCV) (2021)
2. Feng, J.C., Hong, F.T., Zheng, W.S.: MIST: multiple instance self-training framework for video anomaly detection. In: Proceedings of the IEEE/CVF Conference on Computer Vision and Pattern Recognition (CVPR), pp. 14009–14018 (2021)
3. Wu, J.-C., Hsieh, H.-Y., Chen, D.-J., Fuh, C.-S., Liu, T.-L.: Self-supervised sparse representation for Video Anomaly Detection. In: Avidan, S., Brostow, G., Cissé, M., Farinella, G.M., Hassner, T. (eds.) ECCV 2022. LNCS, vol. 13673, pp. 729–745. Springer, Cham (2022). https://doi.org/10.1007/978-3-031-19778-9_42
4. Sultani, W., Chen, C., Shah, M.: Real-world anomaly detection in surveillance videos. In: CVPR, pp. 6479–6488 (2018)
5. Liu, W., Luo, W., Lian, D., Gao, S.: Future frame prediction for anomaly detection–a new baseline. In: CVPR, pp. 6536–6545 (2018)
6. Wu, P., et al.: Not only look, but also listen: learning multimodal violence detection under weak supervision. In: Vedaldi, A., Bischof, H., Brox, T., Frahm, J.M. (eds.) ECCV 2020. LNCS, vol. 12375, pp. 322–339. Springer, Cham (2020). https://doi.org/10.1007/978-3-030-58577-8_20
7. The Digital Research Alliance of Canada (the Alliance), advanced research computing network. https://alliancecan.ca. Accessed November 2023

# Detecting Abnormal Authentication Delays In Identity And Access Management Using Machine Learning

Jiahui Xiang, Osman Salem$^{(\boxtimes)}$, and Ahmed Mehaoua

Centre Borelli, UMR 9010 Université Paris Cité, Paris, France
{jiahui.xiang,osman.salem,ahmed.mehoua}@u-paris.fr

**Abstract.** Authentication delay is an important metric for measuring the performance and responsiveness of Identity and Access Management (IAM) systems. A sudden increase in authentication delay can indicate several problems, such as performance degradation, denial-of-service attacks, or compromised accounts. This paper proposes an adaptive approach for anomaly detection in authentication delay for IAM systems. The proposed approach combines the cumulative sum (CUSUM) algorithm, a statistical method for detecting changes in the mean of a time series, with a Machine Learning (ML) classifier model. The CUSUM algorithm is used to identify potential change points in the authentication delay data and to drive labeled training data, and the ML classifier model is updated using the derived data set for real-time data classification. The proposed approach is adaptive, meaning that it can automatically adjust to changes in the underlying authentication delay distribution. Our experimental results showed that our approach improves the detection accuracy in real-time deployment scenarios.

**Keywords:** Identity and Access Management · Authentication delay · Change Point Detection · Classification models · Cumulative Sum · Okta delegated authentication

## 1 Introduction

Identity and Access Management (IAM) systems are increasingly being deployed in IT environments. In the past decade, IAM has become a critical component of modern security infrastructure, ensuring the availability, confidentiality, and integrity of corporate assets.

Historically, directories were used to manage most access to internal applications, with LDAP [4] being the most common implementation. However, Active Directory (AD [1]) by Microsoft is now the most widely used LDAP implementation, due to its simplicity and scalability. However, AD is vulnerable in multiple ways [2], with new vulnerabilities being discovered all the time, especially as more assets are added to the network. Taking control of AD is a top priority for

E. Renault et al. (Eds.): MLN 2023, LNCS 14525, pp. 57–71, 2024.
https://doi.org/10.1007/978-3-031-59933-0_5

black hat hackers, and most ransomware attacks are deployed through admin accounts using Group Policy Objects (GPOs).

In the realm of security for large-scale environments, whether cloud-based or on-premises, the Identity and Access Management (IAM) system assumes a pivotal role. This robust system stands as a critical line of defense, capable of mitigating vulnerabilities inherent in Active Directory. Its multifaceted capabilities span a spectrum of vital functions, from managing user identities and their access to resources to enforcing stringent measures that safeguard sensitive data from unauthorized access. Authentication, authorization, and auditing are fundamental components at the heart of every IAM setup, actively working to fortify security measures against potential breaches and unauthorized entry.

IAM systems operate on a comprehensive set of principles that underpin their effectiveness. Leveraging its capacity to provide least privilege access, IAM bolsters the security framework, while simultaneously ensuring compliance with stringent regulatory standards such as GDPR and HIPAA. In addition to regulatory adherence, IAM systems streamline the process of employee onboarding and offboarding, thereby fostering a seamless and secure transition for personnel entering or exiting the organization. Single Sign-On (SSO) capabilities further enhance the security apparatus by offering a unified authentication platform for multiple applications, thus significantly reducing the risks associated with password leaks and unauthorized access attempts.

Furthermore, IAM systems bolster their defenses through the enforcement of stringent password policies and the implementation of multi-factor authentication, thus fortifying the security infrastructure against potential breaches. Emphasizing role-based access control, IAM systems effectively limit the scope of user privileges, thereby ensuring a secure and controlled access environment across various organizational levels. By centralizing user management across multiple systems, IAM systems streamline administrative tasks and enhance the overall efficiency of the security ecosystem.

In addition to its core functions, IAM systems also assume an active role in monitoring user behavior, allowing for proactive measures in the event of potential security threats. Adaptive authentication features further reinforce the security posture, dynamically adjusting security protocols based on risk levels and potential threats. By enabling federation, IAM systems facilitate secure access across different organizations, fostering an environment of collaboration and secure data exchange. Additionally, IAM systems play a critical role in securing APIs and microservices, fortifying the system against potential external threats and vulnerabilities.

In the current landscape, the implementation of an IAM system stands as a fundamental best practice for every security infrastructure. Among the leading entities in this domain, Okta emerges as a distinguished leader, renowned for its expertise in Access Management. Okta's prominence and continuous evolution within the realm of Identity Management reinforce its pivotal position within the security landscape, emphasizing its indispensable role in modern security frameworks.

In the context of ensuring minimal service quality for large-scale environments, two pivotal factors come to the forefront: availability and performance. An essential metric in this regard is the authentication delay time. This metric serves as a critical indicator, measuring the time it takes for a user's final authentication request to initiate and receive a valid token from an identity provider. This specific delay time provides a discrete and tangible measurement that is instrumental in the thorough analysis of both the availability and performance of an Identity and Access Management (IAM) product. It forms the cornerstone for evaluating the system's responsiveness and its ability to consistently provide uninterrupted services.

The primary objective of this research paper centers on the identification and mitigation of abnormal authentication delay times within IAM systems. Numerous variables and parameters can contribute to prolonged authentication delays. These factors span a wide spectrum, including network-related elements such as proxies and VPNs, as well as system-level factors like CPU performance (including aspects like endpoint detection and response, and excessive concurrent request loads). The identification of these contributing elements is paramount, as it forms the foundation for optimizing and enhancing the authentication process to ensure minimal delays and an optimal user experience.

The remainder of this paper is structured as follows. Section 2 provides a review of recent work related to the detection of abnormal authentication delays in IAM. In Sect. 3, we introduce our proposed approach, which leverages a change point detection algorithm to generate labeled data and subsequently updates the classification model using the derived labeled data. In Sect. 4, we present the experimental results of our approach and compare the performance of four supervised Machine Learning (ML) methods in terms of their detection accuracy. Finally, Sect. 5 concludes the paper and discusses potential future research directions.

## 2  Related Work

Change point detection in data analysis is a technique specifically crafted to pinpoint shifts within time series data. Its utility extends across diverse domains, including finance, quality control, and anomaly detection. By employing statistical methods, change point detection reveals noteworthy modifications in the underlying data structure, offering valuable insights into these alterations. This process assumes a critical role in proactive decision-making and early problem prevention.

Truong *et al.* in [10] offer a comprehensive analysis of offline change point detection methods, shedding light on their strengths and limitations. This review delves into the techniques employed for identifying structural changes in time-series data, with a particular focus on their applicability across diverse domains. Emphasizing the significance of model and parameter selection in achieving precise change point detection, the paper covers classical methods like CUSUM and Bayesian techniques, as well as recent advancements such as deep learning and online learning algorithms. The computational complexity and scalability of these methods are crucial considerations for real-world applications. The review also highlights emerging trends and points towards potential research directions.

Furthermore, Truong *et al.* in [10] introduce an innovative approach known as Penalty Learning for Changepoint Detection, which significantly enhances the accuracy of detecting structural changes within time-series data. This innovative approach optimizes the penalty term within a cost function to improve changepoint localization, resulting in more precise results. Notably, it offers adaptability to complex data, providing a flexible approach to changepoint detection.

Liu *et al.* in [6] have introduced a pioneering approach for change point detection through relative density-ratio estimation. This innovative method is designed to identify significant shifts in the underlying data distribution, thus enabling the detection of structural abrupt changes. It proves particularly effective for scenarios involving non-stationary and intricate data patterns, offering both precision and computational efficiency in real-time applications.

Machida *et al.* in [5] introduce an innovative and sophisticated approach for detecting change points. This method distinguishes itself by employing cutting-edge subspace identification techniques, which offer a more nuanced understanding of structural changes within multivariate time series data. By leveraging the concept of subspace modeling, the approach enhances our ability to capture and interpret intricate variations in the data, making it particularly valuable for tasks requiring a comprehensive analysis of temporal data patterns.

The field of attack detection in IAM is currently witnessing a significant upsurge in interest and research. Historically, the predominant approach for attack detection has relied on rule-based methods, but with the increasing sophistication of cyberattacks, these traditional techniques have proven insufficient for safeguarding critical systems effectively. ML has emerged as a formidable ally in bolstering IAM security by enhancing attack detection.

Numerous researchers have delved into leveraging ML to fortify IAM systems, devising robust, adaptive, and highly effective defense mechanisms against a myriad of severe security threats. Their work contributes to the evolving landscape of IAM security, as it embraces the power of ML to provide a more resilient and proactive response to the ever-evolving cyber threat landscape.

Authentication is a cornerstone of security within the realm of System and Information Security. However, the landscape of authentication is continually evolving, marked by the increasing challenges posed by sophisticated attacks and erratic user behavior. In response to these challenges, ML has risen as a potent and adaptable tool for fortifying the authentication process, enhancing both security and user experience.

Siddiqui *et al.* in [8] delves into an array of user authentication schemes that harness the capabilities of ML. This comprehensive investigation encompasses the deployment of various ML algorithms, spanning the domains of supervised and unsupervised methods, deep learning, and ensemble techniques. The training of these models commences with an eclectic selection of data sources, such as biometric data, user behavior patterns, keystroke dynamics, and physiological characteristics. These rich sources of information serve as the foundation for constructing robust and adaptive authentication models.

The application of these models extends across a diverse spectrum of authentication schemes, including biometrics, behavior-based authentication, and device recognition. Furthermore, these pioneering research findings find utility in a wide array of domains, ranging from securing mobile devices and web services to safeguarding the burgeoning Internet of Things (IoT) ecosystem. As ML continues to evolve and adapt, it not only empowers more secure and user-friendly authentication but also extends its reach across an expanding array of applications and industries, thus contributing significantly to the dynamic field of Information Security (IS).

In the realm of security, two significant studies have harnessed the power of ML to fortify different aspects of safeguarding sensitive information. Pande *et al.* in [7] have directed their attention toward bolstering the defense against Distributed Denial of Service (DDoS) attacks using ML. They've employed the Random Forest algorithm to classify network traffic as normal or indicative of an attack based on sample data. Impressively, these samples encompass instances of notorious attacks, including the notorious ping of death. The results are remarkable, with an astonishing accuracy rate of 99.76% in classifying these samples.

On a parallel front, Djosic *et al.* in [3] have ventured into enhancing Identity and Access Management (IAM) security through the development of a Risk Authentication Decision Engine. This innovative engine leverages ML algorithms to assess access requests in real-time, enabling it to make well-informed authentication decisions based on a multitude of risk factors. It takes into account user behavior, device attributes, geographical location, and historical data to provide a holistic risk assessment.

Central to the engine's functionality are the core ML techniques, which facilitate risk evaluation and decision-making. The engine is fueled with a diverse array of data sources, including user profiles, access logs, and external threat intelligence, creating a comprehensive foundation for robust security measures. Striking a balance between stringent security protocols and a seamless user experience is a critical aspect of the engine's design, ensuring a smooth and secure authentication process. This Risk Authentication Decision Engine represents an innovative and dynamic solution that embodies the evolution of IAM security through the integration of ML.

## 3    Proposed Approach

In a real System and Information Security (SI) environment, the absence of annotated datasets poses a challenge for machine learning-based intrusion detection. However, several raw factors can be collected, such as authentication delay, datetime, IP addresses, username IDs, browser agents, and host operating systems. To facilitate the development of ML models for intrusion detection, a method for annotating raw data is required. Once this data is annotated, it becomes a valuable resource for training and evaluating machine learning models.

One statistical algorithm that is particularly useful for detecting abrupt changes in data over time is the Cumulative Sum (CUSUM) algorithm [9]. The CUSUM algorithm calculates a CUSUM value for each new data point, with the formula:

$$S_0 = 0 \tag{1}$$
$$S_{n+1} = \max(0, S_n + x_{n+1} - \alpha \times w_n) \tag{2}$$

With $S_n$ is the cumulative sum at the $n$-th iteration, $x_n$ is the measured authentication delay, $alpha$ is a parameter greater than 1, and $w_n$ is the average of the past value of the authentication delay. These CUSUM values are continuously updated as new data points are observed. When the CUSUM value surpasses a predefined threshold $h$, indicating a significant deviation, an alarm is triggered. Notably, the larger the deviation, the faster the CUSUM value accumulates. Following the alarm, the CUSUM value is reset to zero to commence monitoring for the next change. The choice of appropriate values for the parameters $\alpha$ and $h$ is crucial to ensure the algorithm's effectiveness for a specific application. Each alarm raised serves as an annotation, flagging the corresponding data point as an abnormal authentication delay.

In the initial stages, models are trained using pre-existing annotated data. However, in real-world applications, data traffic undergoes constant and dynamic transformations. This is primarily due to the evolution of architecture at various levels, including network configurations, workstation setups, server upgrades, and the growing user base, among other factors. Consequently, the trained model must exhibit adaptability to seamlessly accommodate each incoming stream of new data. The central focus of this paper lies in the heart of this challenge: the development of methodologies to ensure the adaptability of ML models.

We suggest a novel approach that merges the CUSUM algorithm with an ML model. In our methodology, the CUSUM algorithm is employed to label incoming data, and subsequently, this annotated data is utilized for incremental model updates.

In specific scenarios where no new alarms are triggered, there is no necessity to update the ML model. The annotated data is partitioned into training and testing sets using an 80/20 split ratio. The ML models considered in our study encompass Random Forest (RF), K-Nearest Neighbors (KNN), Decision Trees (DT), and Support Vector Machines (SVM).

RF is a versatile and potent machine learning algorithm adept at handling both classification and regression tasks. It belongs to the ensemble learning family, a technique that amalgamates the predictions from numerous decision trees to yield more precise and resilient predictions.

In RF classifier, a substantial number of DTs are individually trained on random subsets of the training data. Each tree autonomously makes predictions, and the final prediction is computed by amalgamating these individual tree predictions. In classification tasks, this aggregation typically involves majority voting, while in regression tasks, it entails averaging the predictions. This ensemble approach serves a dual purpose: it mitigates overfitting and augments the model's ability to generalize well to new data.

RFs are acclaimed for their competence in dealing with high-dimensional datasets, their knack for discerning important features, and their capability to provide insights into the relative significance of variables within the dataset. Consequently, they find extensive application across various domains, including finance, healthcare, and natural language processing, rendering them a favored choice among data scientists and machine learning practitioners.

KNN stands out as a versatile and widely adopted ML algorithm, holding a pivotal role in both classification and regression tasks. What sets KNN apart is its non-parametric and instance-based nature, as it refrains from making extensive assumptions about the underlying data distribution. Instead, it operates on the fundamental principle of proximity.

KNN's modus operandi involves making predictions for unseen data points based on the classes of their nearest neighbors within the training dataset. The parameter "K" plays a critical role in this process, as it specifies the number of nearest neighbors to consider and significantly impacts the algorithm's performance. To gauge this proximity, KNN relies on distance metrics, with the Euclidean distance being a frequently employed measure. Nonetheless, it's imperative to properly scale and preprocess the data to ensure the algorithm's robustness.

KNN's simplicity is juxtaposed with its computational demands, necessitating the calculation of distances to all training data points during prediction. This algorithm finds applications across a broad spectrum of fields, including recommendation systems, image recognition, and anomaly detection. It serves as an effective and intuitive tool for data analysis and decision-making, particularly in scenarios featuring well-defined class boundaries and moderately sized datasets.

A DT is a foundational and interpretable ML algorithm that finds utility in both classification and regression tasks. It takes the form of a flowchart-like structure, where each node signifies a feature or attribute, and each branch represents a potential outcome or decision contingent on the feature's value. At the terminal nodes, you'll find the ultimate predictions or class labels.

What distinguishes DT is their innate simplicity and transparency, making them exceptionally valuable for elucidating the rationale behind a model's predictions. These trees frequently serve as the building blocks for more intricate ensemble methods such as Random Forests and Gradient Boosting.

One of their key strengths is their ability to efficiently manage both categorical and numerical data, automatically identifying the most crucial features for decision-making. This adaptability renders DT a versatile tool in the realm of ML. Nonetheless, they are susceptible to overfitting, a challenge that can be addressed through pruning techniques or by incorporating ensemble methods.

SVM possesses a unique ability to identify an optimal hyperplane that maximizes the margin between different classes within a dataset. This hyperplane serves as the decision boundary, and the support vectors, which are the data points closest to it, play a crucial role in determining its placement. SVMs are particularly valuable when tackling intricate, high-dimensional data and exhibit versatility in handling both linear and nonlinear classification challenges, thanks to the utilization of kernel functions.

One of the standout features of SVMs is their adaptability and capacity to generalize effectively, even when the data lacks linear separability. Their optimization process is geared toward maximizing the margin while minimizing classification errors, which, in turn, aids in mitigating overfitting.

SVMs are employed across a wide spectrum of fields, encompassing image recognition, text classification, and bioinformatics, due to their resilience, strong theoretical foundation, and versatility in managing diverse data types. However, it's worth noting that SVMs can be computationally intensive for large datasets, and careful parameter tuning is essential for their success. When applied thoughtfully, they consistently yield exceptional results in numerous real-world applications.

## 4   Experimental Results

The IAM architecture experimented with in this paper is as follows: Okta, a SAS product offering advanced authentication services, is coupled with Active Directory, which manages identities and passwords, through an Okta AD agent. Both agents and workstations communicate with Okta through a Zscaler proxy. Workstations may or may not use a VPN to connect to internal IT networks.

In this paper, 59 domains are synchronized by at least two agents each to a single instance of Okta. Figure 1 illustrates Okta Delegated Authentication for one domain.

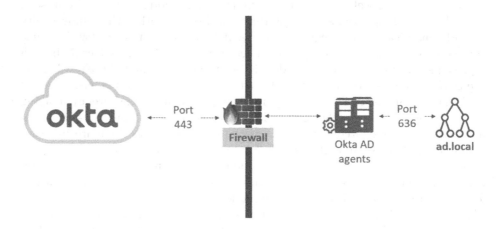

**Fig. 1.** Okta - AD: Delegated authentication architecture

All datasets were collected from Okta logs from September 2022 to April 2023 and contain 63 Active Directory domains. The logs were in JSON format, which we extracted and parsed into CSV format for each domain. We only used the authentication delay time and the time the log was registered. The domain names were anonymized. All experiments were performed on the following machine:

- Machine: T480 (2020)
- CPU: i7-8550U (4 cores - 8 threads/min 1.8 GHz - max 4.0 GHz)
- Memory: 24 GB RAM

The authentication delay was annotated as normal or abnormal, as shown in Table 1. CUSUM was executed on each of the 59 collected domains. This resulted in Table 2, which shows the number of logs, normal authentications, and abnormal authentications for each domain.

**Table 1.** Class Identification Index

| Number | Category |
|--------|----------|
| 0 | Normal |
| 1 | Abnormal |

CUSUM was implemented with a threshold of 900, chosen by heuristic search. Only edge values were selected. CUSUM was tested on each domain of varying size to obtain an indicator of execution speed, as shown in Table 3. On domains with fewer than 200,000 rows, CUSUM detected all alarms in less than 1 s. On domains with between 250,000 and 500,000 rows, it took 2 s. On domains with 1,000,000 rows, it took 4 s. As the data for each domain will only increase over time, the contribution of machine learning will be beneficial in accelerating detection time.

The performance of the ML models was only measured on the 7 domains with the most logs, which ranged from 264,460 to 1,139,570 lines. SVM had the worst performance, with a training time of 18 minutes for domains with 1 million rows. RF came next, with a training time of 10 s. KNN and DT had fairly similar training times. Table 4 and Fig. 2 shows the results.

However, the prediction times for the ML models were much faster. On domains with 1 million rows, KNN predicted in 30 s, RF in 1 s, SVM in just under a second, and DT in around 0.01 s. Table 5 and Fig. 3 shows the prediction times on the measured domains.

Table 6 shows the mean performance time for the machine learning models. On average, DT trains and predicts in the order of 0.001 s.

Models are evaluated based on their performance in terms of accuracy, F1-score, recall, precision, Receiver Operating Characteristic (ROC) curve, and Area Under the Curve (AUC), as well as their time to train and predict. These metrics are defined as follows:

**Table 2.** Dataset

| Samples | 0 | 1 | Samples | 0 | 1 |
|---|---|---|---|---|---|
| 87465 | 87016 | 449 | 65478 | 65464 | 14 |
| 41849 | 41828 | 21 | 240430 | 240288 | 142 |
| 21018 | 21004 | 14 | 112201 | 112156 | 45 |
| 264460 | 264348 | 112 | 410047 | 409475 | 572 |
| 1813 | 1809 | 4 | 6309 | 6302 | 7 |
| 42916 | 42798 | 118 | 51046 | 51043 | 3 |
| 190 | 190 | 0 | 55115 | 55113 | 2 |
| 79252 | 79055 | 197 | 1130570 | 1130345 | 225 |
| 179136 | 179077 | 59 | 100539 | 100466 | 73 |
| 33444 | 33438 | 6 | 6256 | 6254 | 2 |
| 5853 | 5851 | 2 | 117714 | 117594 | 120 |
| 3639 | 3634 | 5 | 975977 | 975388 | 589 |
| 6230 | 6223 | 7 | 318 | 316 | 2 |
| 3728 | 3728 | 0 | 157635 | 157619 | 16 |
| 151923 | 151894 | 29 | 314 | 314 | 0 |
| 6447 | 6436 | 11 | 74392 | 74364 | 28 |
| 332 | 330 | 2 | 61276 | 61263 | 13 |
| 63889 | 63887 | 2 | 112 | 112 | 0 |
| 340037 | 340002 | 35 | 208883 | 208851 | 32 |
| 15106 | 15104 | 2 | 35924 | 35918 | 6 |
| 580980 | 580830 | 150 | 4301 | 4301 | 0 |
| 7727 | 7719 | 8 | 1301 | 1301 | 0 |
| 300517 | 300468 | 49 | 5491 | 5484 | 7 |
| 26 | 26 | 0 | 327404 | 327296 | 108 |
| 23351 | 23349 | 2 | 12149 | 12145 | 4 |
| 1731 | 1731 | 0 | 67693 | 67692 | 1 |
| 1206 | 1202 | 4 | 38355 | 38339 | 16 |
| 4365 | 4365 | 0 | 18696 | 18659 | 37 |
| 20560 | 20560 | 0 | 3266 | 3266 | 0 |
| 46745 | 41014 | 5731 | | | |

**Table 3.** Performance CUSUM on each domain size

| Size | Time (s) | Size | Time (s) |
|------|----------|------|----------|
| 26 | 0.0001747 | 38355 | 0.1547005 |
| 112 | 0.0002592 | 41849 | 0.1752543 |
| 190 | 0.0004168 | 42916 | 0.1782002 |
| 314 | 0.0006409 | 46745 | 0.3606312 |
| 318 | 0.0014043 | 51046 | 0.2189102 |
| 332 | 0.0014563 | 55115 | 0.2217739 |
| 1206 | 0.0055497 | 61276 | 0.2417336 |
| 1301 | 0.0028634 | 63889 | 0.2563026 |
| 1731 | 0.0034647 | 65478 | 0.261126 |
| 1813 | 0.0073297 | 67693 | 0.2681606 |
| 3266 | 0.0067611 | 74392 | 0.2957768 |
| 3639 | 0.015146 | 79252 | 0.3586895 |
| 3728 | 0.0074437 | 87465 | 0.3752143 |
| 4301 | 0.0088849 | 100539 | 0.3954411 |
| 4365 | 0.0094261 | 112201 | 0.4732082 |
| 5491 | 0.0216315 | 117714 | 0.4805808 |
| 5853 | 0.0238254 | 151923 | 0.6101294 |
| 6230 | 0.0255096 | 157635 | 0.6376877 |
| 6256 | 0.0249622 | 179136 | 0.7193756 |
| 6309 | 0.0399289 | 208883 | 0.8764 |
| 6447 | 0.0262887 | 240430 | 0.9879026 |
| 7727 | 0.0320623 | 264460 | 1.0692272 |
| 12149 | 0.0490916 | 300517 | 1.2142997 |
| 15106 | 0.0623329 | 327404 | 1.2777286 |
| 18696 | 0.0805817 | 340037 | 1.3660059 |
| 20560 | 0.0430574 | 410047 | 1.7272482 |
| 21018 | 0.08445 | 580980 | 2.3883655 |
| 23351 | 0.0927732 | 975977 | 3.9321253 |
| 33444 | 0.134294 | 1130570 | 4.630055 |
| 35924 | 0.1484749 | | |

**Table 4.** Performance ML-model to train on domain size in seconde

| Size | RF | KNN | DT | SVM |
|------|------|------|------|------|
| 1130570 | 12.02869 | 0.27505 | 0.08053 | 1135.18603 |
| 975977 | 14.35661 | 0.23037 | 0.14198 | 1135.13952 |
| 580980 | 7.97773 | 0.14261 | 0.09249 | 3142.95517 |
| 410047 | 8.61467 | 0.12907 | 0.08615 | 189.56657 |
| 340037 | 4.06584 | 0.06975 | 0.04878 | 06.50586 |
| 300517 | 2.69473 | 0.07390 | 0.02201 | 253.09265 |
| 264460 | 3.72341 | 0.07326 | 6.94285 | 1.30845 |

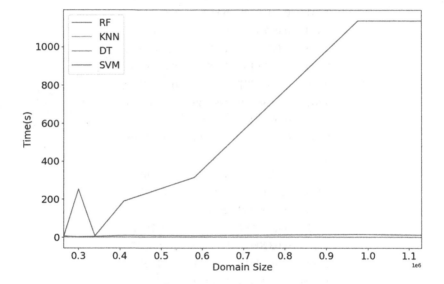

**Fig. 2.** Performance ML-model to train on domain size in seconde

**Table 5.** Performance ML-model to predict on domains size in seconde

| Size | RF | KNN | DT | SVM |
|------|------|------|------|------|
| 1130570 | 1.55665 | 37.78894 | 0.01050 | 0.63847 |
| 975977 | 1.37749 | 30.920587 | 0.00929 | 0.78415 |
| 580980 | 0.92338 | 12.55332 | 0.00586 | 0.21949 |
| 410047 | 0.76587 | 10.28818 | 0.00556 | 0.32098 |
| 340037 | 0.47398 | 6.80992 | 0.00500 | 0.02829 |
| 300517 | 0.54838 | 6.20262 | 0.00288 | 0.04768 |
| 264460 | 0.44821 | 6.94285 | 0.00423 | 0.02413 |

**Table 6.** Model mean performance in time

| Model | Train (s) | Prediction (s) |
|-------|-----------|----------------|
| RF | 3.193194 | 0.379650 |
| KNN | 0.57001 | 5.438721 |
| DT | 0.017129 | 0.002368 |
| SVM | 532.746680 | 0.062506 |

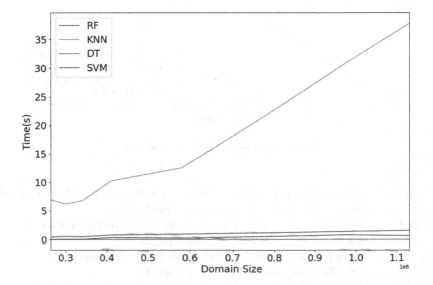

**Fig. 3.** Performance ML-model to predict on domains size in seconde

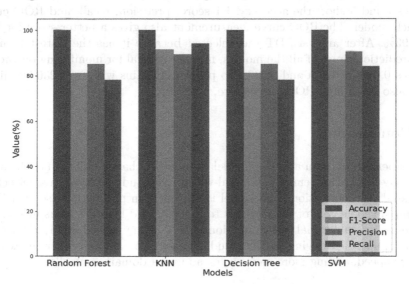

**Fig. 4.** Accuracy, F1-Score, Precision and Recall for Each Tested Model

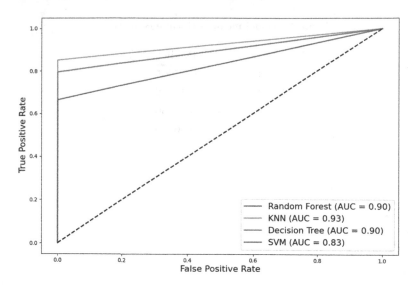

**Fig. 5.** ROC for Each Tested Model

- Accuracy: The proportion of correct predictions to the total predictions made.
- Precision: The proportion of correct positive predictions.
- Recall: The proportion of actual positives that are correctly predicted.
- F1-score: A harmonic mean of precision and recall.
- AUC: A measure of the overall discriminative power of a model.
- Training Time: The time required to train a model.
- Prediction Time: The time required to predict the class of a new data instance.

Figures 4 and 5 show the accuracy, F1 score, precision, recall, and ROC curve for each model. The ROC curve measurement also gives a better score for DT, with 93%. After analysis, DT was selected because it has the fastest training and prediction times of all the models, making it ideal for monitoring on the fly. It takes 0.01 s to train and 0.002 s to predict domains with over 250,000 lines. KNN also has a high ROC curve score, over 90%.

## 5   Conclusion

This paper proposes an adaptive model to monitor change points in authentication delay in IAM systems using a real-world case study: Okta Active Directory Delegated Authentication. The model is currently in production use at a CAC 40 company. The goal of the model is to monitor and help IT teams investigate and remediate abnormal authentication delays. Many remediation actions can be taken, such as moving to the cloud for each domain controller, increasing network speed, adding more memory, or adding a domain controller.

The model can be improved by monitoring additional features, such as identities, network connections, user agents, and geolocation. Other algorithms, such as those implemented in Ruptures, could also be investigated to replace CUSUM as the alarm trigger.

# References

1. Allen, R., Lowe-Norris, A.: Active Directory. O'Reilly Media, Inc., Sebastopol (2003)
2. Binduf, A., Alamoudi, H.O., Balahmar, H., Alshamrani, S., Al-Omar, H., Nagy, N.: Active directory and related aspects of security. In: 2018 21st Saudi Computer Society National Computer Conference (NCC), pp. 4474–4479. IEEE (2018)
3. Djosic, N., Nokovic, B., Sharieh, S.: Machine learning in action: securing IAM API by risk authentication decision engine. In: 2020 IEEE Conference on Communications and Network Security (CNS), pp. 1–4. IEEE (2020)
4. Howes, T., Smith, M., Good, G.S.: Understanding and Deploying LDAP Directory Services. Addison-Wesley Professional, Boston (2003)
5. Kawahara, Y., Yairi, T., Machida, K.: Change-point detection in time-series data based on subspace identification. In: Seventh IEEE International Conference on Data Mining (ICDM 2007), pp. 559–564. IEEE (2007)
6. Liu, S., Yamada, M., Collier, N., Sugiyama, M.: Change-point detection in time-series data by relative density-ratio estimation. Neural Netw. **43**, 72–83 (2013)
7. Pande, S., Khamparia, A., Gupta, D., Thanh, D.N.H.: DDOS detection using machine learning technique. In: Khanna, A., Singh, A.K., Swaroop, A. (eds.) Recent Studies on Computational Intelligence. SCI, vol. 921, pp. 59–68. Springer, Singapore (2021). https://doi.org/10.1007/978-981-15-8469-5_5
8. Siddiqui, N., Pryor, L., Dave, R.: User authentication schemes using machine learning methods—a review. In: Kumar, S., Purohit, S.D., Hiranwal, S., Prasad, M. (eds.) Proceedings of International Conference on Communication and Computational Technologies. AIS, pp. 703–723. Springer, Singapore (2021). https://doi.org/10.1007/978-981-16-3246-4_54
9. Siris, V.A., Papagalou, F.: Application of anomaly detection algorithms for detecting SYN flooding attacks. In: IEEE Global Telecommunications Conference, 2004. GLOBECOM'04, vol. 4, pp. 2050–2054. IEEE (2004)
10. Truong, C., Oudre, L., Vayatis, N.: Selective review of offline change point detection methods. Signal Process. **167**, 107299 (2020)

# SIP-DDoS: SIP Framework for DDoS Intrusion Detection Based on Recurrent Neural Networks

Oussama Sbai$^{(\boxtimes)}$ ⓘ, Benjamin Allaert ⓘ, Patrick Sondi ⓘ,
and Ahmed Meddahi ⓘ

Centre for Digital Systems, IMT Nord Europe, Institut Mines-Télécom,
59000 Lille, France
{oussama.sbai,benjamin.allaert,patrick.sondi,
ahmed.meddahi}@imt-nord-europe.fr

**Abstract.** The rapid evolution of beyond fifth-generation (B5G) and sixth-generation (6G) networks has significantly driven the growth of Internet of Things (IoT) applications. These applications are characterised by: a massive connectivity, high security level, trust, wireless coverage, also ultra-low latency, high throughput, and ultra-reliability, especially for real-time oriented sessions or sensor like cameras. While traditional protocols like MQTT and CoAP are inadequate for such types of applications, under certain conditions, the 3GPP standard Session Initiation Protocol (SIP) emerges as a promising solution. However, SIP faces various Distributed Denial of Service (DDoS) threats, as INVITE flooding attacks presenting a significant challenge. This work presents a GRU-based Intrusion Detection System (IDS) to detect SIP-INVITE flooding attacks. Leveraging recurrent neural networks, the IDS efficiently process sequential SIP traffic data in real time, identifying attack patterns effectively. The GRU's ability to capture temporal dependencies enhances accuracy in classifying and detecting attack behaviors. The results demonstrate that the framework can effectively detect and mitigate INVITE flooding attacks of different intensities, under practical settings. The performance results show that the proposed framework is robust and can be practically deployed, e.g., inference time less than 800 μs and a marginal rate for the misclassified traffic.

**Keywords:** SIP protocol · DoS/DDoS · SIP-INVITE flooding attack · Recurrent Neural Networks · IoT

## 1 Introduction

The Internet of Things (IoT) has become a powerful and transformative force, driven by the goal to improve safety, efficiency, and sustainability for a wide range of applications. IoT is one of the most promising technological advancements on the horizon is the integration of sixth-generation (6G) networks and the evolution beyond 5G (B5G) as it has proven capacity for innovation and enhancement.

E. Renault et al. (Eds.): MLN 2023, LNCS 14525, pp. 72–89, 2024.
https://doi.org/10.1007/978-3-031-59933-0_6

This development has the potential to revolutionize communication systems by optimizing processes within smart cities, industries, or powering the seamless operation of intelligent transportation systems (ITS). This also by introducing high speeds, ultra-low latency, and a vastly expanded capacity. These attributes are indispensable for accommodating the expected exponential growth in connectivity. The convergence of the Internet of Things (IoT) with 6G and B5G networks presents an unprecedented opportunity to transform the digital landscape. This technologies integration will enable the deployment of innovative IoT applications that require real-time constraints, reliable communication channels, efficient and predictive resource management, and robust security systems.

In this context, the network communications require "lightweight" protocols for session initiation and control, such as Message Queuing Telemetry Transport (MQTT) and Constrained Application Protocol (CoAP). However, MQTT and CoAP are not applicable in advanced IoT applications that require high data rates and low latency [22,23]. Session Initiation Protocol (SIP) is considered as a good candidate, for supporting a wide variety of IoT application scenarios [7,11,22,23]. Besides instant messaging, it is suitable for handling short or long session semantics for streaming data and publish-subscribe semantics for event notifications, e.g., subscribe, publish, notify or method, thereby enabling more sophisticated IoT network operations. The Session Initiation Protocol (SIP) is well-suited for this purpose and functions as a signaling protocol designed for multimedia communication [9,22,23]. Nevertheless, one of the major SIP Distributed Denial of Service (DDoS) attacks is the INVITE flooding attack [5]. In this attack, malicious users flood the proxy or SIP server with a high volume of malicious INVITE messages with the aim of disrupting the service.

In this paper, we propose a comprehensive framework specifically designed to detect SIP INVITE flooding attacks. Recent approaches to detecting SIP-DDoS attacks focus mainly on analyzing the content of SIP messages or the entire SIP dialog [14,16]. However, these works have not been studied to analyze SIP traffic in time intervals. This temporal segmentation makes it possible to systematically examine and evaluate SIP communication patterns, while facilitating timely anomaly detection and improving the overall effectiveness of the IDS. The proposed framework is designed to address the unique challenges posed by such attacks, and to provide an effective defense mechanism against DoS/DDoS threats targeting SIP infrastructures. The core of the proposed framework is a Gated Recurrent Unit (GRU) based Intrusion Detection System (IDS). By harnessing the power of recurrent neural networks, the IDS can effectively process sequential SIP traffic data and identify patterns indicative of INVITE flooding attacks. The GRU's ability to capture temporal dependencies makes it a valuable asset in recognizing attack behaviors with high accuracy. To evaluate the proposed IDS, the Train/Test/Validation protocol was employed, a widely-used technique for assessing the model's performance. Furthermore, an independent test dataset was used to validate the IDS's effectiveness. The results demonstrate the framework's ability to accurately detect and mitigate INVITE flooding attacks, even under different attack intensities, showcasing its robustness and practicality in real-world settings. The proposed framework is available online[1] containing the code, dataset and learning models. The main contributions are the followings:

– Generating a realistic dataset: to ensure the effectiveness of the framework, a diverse and realistic dataset was generated using the SIPp [19] and Mr.SIP [21] tools. This dataset contains a legitimate and malicious INVITE messages with different scenarios and message rates, providing a solid basis for training and evaluating our proposed deep learning based IDS.
– Proposing a reliable and reproducible evaluation protocol: a strategy for encoding SIP message frames is proposed to facilitate learning model convergence. Several recurrent neural network models are confronted in order to study their ability to detect anomalies in different attack scenarios. A study of detection accuracy and inference time is carried out.

The paper is structured as follows. Section 2 reviews the related work, discussing existing research in the field. Section 3 presents the proposed framework, detailing the experimental platform, the dataset generation process, and the approach for detecting SIP DDoS attacks. Section 4 describes the detection methodology, and Sect. 5 presents the results obtained from deploying the proposed approach. Finally, we conclude by highlighting the key findings and suggesting directions for future research.

## 2   Related Works

The related literature shows that DDoS attacks can be considered as one of the most predominant threats in IoT networks [6,8,12,15]. DDoS attacks pose significant challenges to the security and stability of IoT infrastructures due to their ability to disrupt services by overwhelming target systems with a massive flow of malicious traffic. This section provides an overview of the proposed dataset for SIP-DoS/DDoS transactions, as well as a review of various studies that address approaches to detecting SIP anomalies in terms of data encoding and technologies used.

### 2.1   SIP Dataset

Few datasets have been proposed in the literature for analyzing DDoS attacks. Alvares et al. [1] introduce a dataset dedicated to Voice over Internet Protocol (VoIP) networks, while considering various DoS/DDoS attack types, including the INVITE flooding attack. Nassar et al. [13] present a dataset with multiple SIP-DoS/DDoS attacks, including INVITE flooding. They used "Inviteflood" tool with two intensity scenarios: 100 and 1000 requests per second.

However, these two existing datasets suffer from same limitation or lack of details as they do not provide a comprehensive details regarding the considered

---

[1] Repository: https://gvipers.imt-nord-europe.fr/benjamin.allaert/sip-ddos.

specific attack, especially they do not consider the various types of SIP-INVITE flooding attacks and scenarios. Additionally, they do not address the critical issue of invalid IP addresses, in case of IP spoofing attacks. This information can provide a more holistic and realistic representation of practical intrusion scenarios. These shortcomings tend to reduce the ability to properly evaluate IDSs for SIP-INVITE attacks.

## 2.2 SIP Anomalies Detection

Recent approaches to detecting SIP-DDoS attacks focus mainly on analyzing the content of SIP messages or the entire SIP dialog. Pereira et al. [17] present an IDS model for recognizing SIP signaling patterns to identify abnormal SIP dialogs within observed SIP sequences. Their core approach is based on a LSTM architecture. They compared their methods to a probabilistic-based solution and found that their methods achieved higher detection scores in a shorter time. Extension [16] have been proposed for signaling SIP attacks based on Convolutional Neural Networks (CNN) architecture. The performance evaluation demonstrates that both the CNN and LSTM models achieve similar effectiveness in detecting the most probable SIP dialog identifier. Nazih et al. [14] introduce an IDS based on the RNN architecture, designed to analyze message content and identify SIP-INVITE flooding attacks. To evaluate the effectiveness of their proposed solution, the authors conduct experiments using a dataset that includes real legitimate SIP messages (normal usage), while the malicious messages (INVITE flooding attacks) are generated using the SIPp-DD tool [20]. However, these works have not been studied to analyze SIP traffic in time intervals. This temporal segmentation makes it possible to systematically examine and evaluate SIP communication patterns, while facilitating timely anomaly detection and improving the overall effectiveness of the IDS.

Several SIP message encoding and artificial data generation techniques have been proposed to improve the ability of neural networks to detect DDoS attacks. Nazih et al. [14] provide a character- or token-based feature extraction process. Raw data is transformed into sequences, following by padding techniques to ensure consistent sequence lengths. Embedding techniques are then employed to represent the data into a format that can be exploited for a meaningful analysis and pattern recognition, related to the SIP-INVITE message. The results clearly demonstrate that the token-based approach combined with the GRU and LSTM architectures give the best performance. Meddahi et al. [10] introduce Generative Adversarial Networks (GAN) to augment SIP messages. The Authors convert the SIP traffic data into images, enabling image-based techniques to be applied to SIP traffic data. The final phase involves the deployment of a GAN model utilized to generate new SIP messages. These newly generated SIP messages are then integrated into the dataset, expanding its size and diversity, effectively. The authors introduce a parameter $\gamma$ to measure the gap between the synthetic and real SIP-data. Nevertheless, data augmentation using techniques like GAN introduce information loss in the SIP message fields. This limitation is critical for message and traffic analysis in case of intrusion detection. While these techniques have proved effective for detecting SIP-DDoS attacks, focusing on the analysis of SIP message content or the SIP dialog itself, they have not been proven effective for analyzing an attack in real-time sequential SIP traffic data.

## 3   Framework

This section, is dedicated to describe the proposed framework. We provide a comprehensive overview of the SIP Protocol, its core principles, and function-alities. We also introduce the tools used in our experimentation, including the simulation environment and data generation techniques. Finally, we outline our SIP dataset, with a focus on the different SIP-INVITE flooding attack scenarios.

### 3.1   SIP Protocol Overview

The Session Initiation Protocol (SIP) stands as a foundational signaling proto-col within the domain of real-time communication [RFC3261-3265], primarily designed to initiate, maintain, modify, and terminate communication sessions between Internet Protocol (IP) devices. SIP finds application in a diverse array of scenarios, encompassing Voice over IP (VoIP), video conferencing, instant messaging, and presence services.

SIP adopts a text-based protocol architecture, following a request-response paradigm. Typically transmitted over User Datagram Protocol (UDP), although Transmission Control Protocol (TCP) is also viable, SIP messages are structured with a header-body format. The header comprises essential information concern-ing message attributes, source and destination addresses, and various parame-ters, while the body carries supplementary data, such as session descriptions or multimedia streams.

SIP session establishment is initiated through the transmission of an INVITE message from the caller to the callee. The INVITE message includes a session description, detailing the caller's preferred media types and supported codecs. The callee, upon accepting the session, responds with a 200 OK message, marking the establishment of the session. Subsequently, participants can exchange media streams using the designated codecs.

SIP allows for session modifications and terminations via corresponding SIP messages. To adjust a session, participants may employ a REINVITE message, while session termination is accomplished through a BYE message. The detailed process of the SIP call is illustrated in Fig. 1.

Distinguished by its adaptability and extensibility, SIP incorporates various noteworthy features:

- Scalability: SIP is architected to scale efficiently, accommodating a substantial user base and numerous concurrent sessions;
- Reliability: SIP ensures message reliability through mechanisms designed to guarantee successful delivery;
- Security: SIP boasts a range of security mechanisms, encompassing authen-tication and encryption;

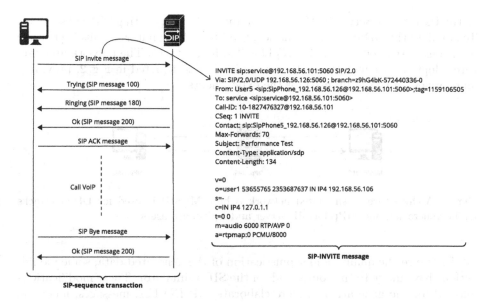

**Fig. 1.** An overview of SIP session transaction and an example of SIP-INVITE message

– Mobility: SIP supports user mobility, allowing seamless session transition as users move between locations.

Widely adopted in the telecommunications sector, the SIP protocol is the backbone of VoIP and real-time communication services offered by many providers. It also finds extensive application in enterprise solutions, including video conferencing and instant messaging.

Several real-world instances underscore SIP's ubiquity:

– VoIP Calls: SIP underpins VoIP calls via mobile applications, orchestrating call initiation and management;
– Video Conferencing: SIP facilitates the establishment and control of video conferences, whether through web browsers or dedicated conferencing clients;
– Instant Messaging: In messaging applications, SIP may be leveraged for message delivery;
– Presence Services: SIP plays a role in delivering presence information, informing users about the online status of their contacts.

SIP stands as a versatile and influential protocol underpinning a broad spectrum of real-time communication applications, exemplifying its indispensable role within contemporary telecommunications infrastructure.

## 3.2   SIP Communication Network Simulation

The proposed dataset was generated using the SIPp simulator, which served as the User Agent Client (UAC) for generating legitimate INVITE messages, and as the User Agent Server (UAS) for receiving and manipulating SIP call sessions. To simulate the DDoS attack scenarios, the Mr.SIP simulator was utilized as the malicious UAC, generating the INVITE flooding attacks. The network architecture adopted for simulating SIP communication is depicted in Fig. 2, providing an overview of the setup used for the experiments.

Mr.SIP: UAC                SIPp: UAS                SIPp: UAC

**Fig. 2.** Architecture of simulated network, where Mr.SIP is used for DDoS attacks traffic generation and SIPp for SIP server and legitimate users

To ensure the quality and sophistication of the generated data, some modifications have made to the source code of the SIPp simulator. These modifications allowed for the generation of more elaborate SIP INVITE messages, including the main SIP fields such as the source IP address, branch number, URI, and call number. Figure 3 shows that SIPp is not full compliant with SIP specifications as the fields of the original SIP INVITE message are not randomly generated. This fields are based on a specific format that does not comply with the SIP specifications RFC 3261 [18]. As a result, the generated SIPp messages deviated from the conventional format. The same figure shows an example of the new SIP message generated by SIPp, after the source code modifications were made to enhance and modify the SIP message structure. Furthermore, in addition to generating legitimate INVITE messages, the Mr.SIP tool plays a crucial role in simulating multiple attackers and generating DoS/DDoS attacks through IP-spoofing techniques. The ability to mimic multiple attackers allows us to create realistic and challenging scenarios, simulating the presence of malicious actors attempting to overwhelm the SIP infrastructure with a barrage of malicious INVITE messages.

### 3.3    Training SIP Data Generation

The INVITE flooding attack is executed by an attacker who generates malicious INVITE messages, by spoofing the IP address and/or URI of legitimate users. The primary objective of this attack is to overwhelm the SIP server by not responding to the SIP 200 message with an ACK SIP message. Consequently, the server enters an ACK wait phase (timeout) while attempting to reestablish communication with the client through a SIP 200 message. The attack's effectiveness lies in the large volume of INVITE messages sent without corresponding ACK messages, leading to a denial of service for the SIP server.

The Mr.SIP tool uses IP spoofing attack to hide the true source of its malicious traffic, making it difficult for the target system to identify the attacker and detect the attack. This enables to evaluate the proposed framework's robustness

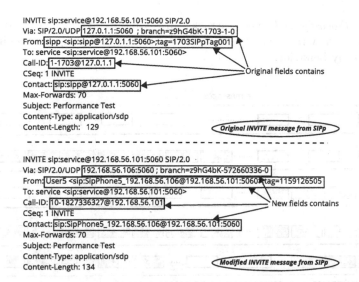

**Fig. 3.** Changes made to the content of INVITE message fields generated by SIPp to ensure the consistency of the generated data.

against sophisticated DDoS attacks that try to obfuscate their origins, mirroring real-world threat scenarios.

In the proposed dataset, the focus is on capturing the root cause of the attacks, which is the excessive volume of INVITE messages without ACK messages. To highlight this aspect, only the SIP transaction of the INVITE message is saved, omitting other related messages. This dataset structure allows for a more specific analysis of the attack pattern. From the literature [5], on the SIP-INVITE flooding rate, the attack can be classified flowing four categories:

*Abrupt Flooding Behavior:* the most common type of flooding attack in SIP networks. They occur when the SIP proxy or server receives a large number of requests in a short period of time, in an unusual way. This suggests that the attack does not have a specific rate or pattern.

*Very High Flooding Behavior:* the goal of this attack is to overload the system with a large number of INVITE requests in a short space of time. The system may then become unavailable or crash.

*Stealthy Flooding Behavior:* designed to evade detection by IDS. Attackers launch stealthy flooding attacks at a slow rate in order to avoid triggering any alarms or alerts. The low rate of this form of attack makes it extremely difficult to distinguish it from regular behavior, and it keeps the flooding rate slightly below the threshold.

*Low Rate Flooding Behavior:* attacks involving floods can occasionally be launched at a slow rate. The low rate attack has roughly constant flooding rate, unlike stealthy behavior.

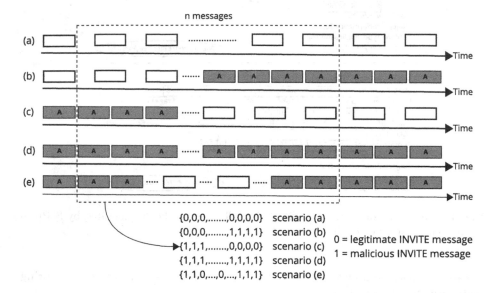

**Fig. 4.** The different generated sequence scenarios of INVITE traffic. (a) legitimate sequence, (b) a sequence that ends with malicious traffic, (c) a sequence that starts with malicious traffic, (d) a sequence containing nothing but malicious traffic, (e) a sequence containing legitimate traffic in the middle.

Giving this, the dataset contains various sequences and scenarios of the INVITE flooding attack, as depicted in Fig. 4. Additionally, it includes different sub-classes of attacks, providing further granularity and insights:

- Very low (VL): 10 messages/sec
- Low (L): 25 messages/sec
- Medium (M): 50 messages/sec
- High (H): 100 messages/sec
- Very High (VH): 500 messages/sec
- Very High ++ (VH++): 1,000 messages/sec

## 4   SIP-INVITE Flooding Detection

First, the methodology is described, including the sliding window technique and its specific parameters, the datasets distribution and the validation process designed to rigorously evaluate the proposed IDS. In addition, we discuss the selected features for our training model and describe the deep learning algorithms used in our IDS. The effectiveness of our approach is measured, thus the performance metrics we used to evaluate the model's classification of malicious and legitimate SIP-INVITE traffic are described.

## 4.1  Analyze SIP Traffic in Time Intervals

The proposed approach to detecting malicious sequences employs a sliding window technique with specific window size and forward step values. The sliding window size is fixed to 10 messages, corresponds to the "very Low" scenario rate flooding attack, e.g., 10 messages per second. This window size represents the worst case scenario. The assumption is to train the learning model to detect the "worst case" scenario, so that the model is able to detect other, less discrete attack scenarios. The forward step value $p$ between two successive subsequences of transmission flows to be analyzed is set at 2, representing a shift of 2 messages per model inference, thus ensuring overlap between two successive analyses. This sliding window value is justified as the attacker may send only one malicious INVITE message to discover and identify the SIP server or proxy. However, once the attacker sends more than one malicious INVITE message, it indicates a potential flooding attack. By using a forward step of 2 messages, the proposed approach can effectively capture and identify the transition from potential to an effective flooding behavior. The Fig. 5 describes the used sliding window technique. Furthermore, regarding the encoding of SIP-INVITE sequences, a sequence is labeled as malicious if it contains a least 2 malicious messages. Otherwise, it is a legitimate sequence. This rule is explicitly outlined in Eq. (1).

$$\forall x_i, x_j \in S; \; i \neq j; \;\; \text{if } \exists x_i = 1 \text{ and } x_j = 1 \Rightarrow S = 1 \tag{1}$$

where "x" represents an INVITE message and "S" is a set of 10 INVITE messages representing SIP-INVITE sequence ($S = \{x_0, \ldots, x_9\}$). Each "x" within the set "S" can take one of two values: "0" signifies a legitimate message, while "1" is a malicious message ($x \in \{0, 1\}$).

By configuring the sliding window with these specific parameters, e.g., 10 messages/sec and forward step 2, we enhance the detection capability of the proposed approach to accurately identify and classify malicious sequences in SIP communication. Indeed, the choice of a step of 2 messages can be seen as a pessimistic approach, as it places us in the worst-case scenario. However, this approach has several advantages. First, it allows us to detect attacks more quickly, the second, it is more robust to scenarios with fewer malicious messages in the sliding window.

## 4.2  Dataset Distribution and Validation Protocol

To evaluate the performance of the proposed IDS system, the Train, Validation and Test protocol was adopted with distinct Train/Validation datasets and two

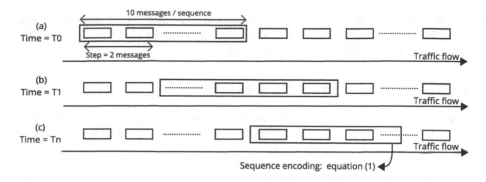

**Fig. 5.** Description of INVITE sequence detection approach: the sliding window size is fixed to 10 messages and the step forward to 2 messages per time unit. If the sequence contains at least 2 malicious messages, it is labeled as malicious. Otherwise, it is a legitimate sequence.

separate Test datasets. This protocol allows to assess the effectiveness of the system across various datasets and ensure its generalisation. The first Test dataset is generated on the same local network used for training. This enables to evaluate the performance of the IDS on typical data, guaranteeing its effectiveness in detecting DDoS attacks in known scenarios. The second Test dataset is generated in a other local network, i.e., the IP address of the SIP client and the SIP server are different from those used in the training phase. This scenario simulates new data (previously unseen), representing real-world type scenarios. Evaluating the IDS on this new data help to asses its generic aspect while detecting DDoS attacks in diverse SIP environments.

The dataset used for training contain 1,135,855 sequences, providing a substantial amount of data for the model. To ensure the effectiveness of the model, a separate validation set was created, consisting of 77,671 sequences. This validation set is used for the evaluation and fine-tuning of the model's performance during the training process. Additionally, a test set comprising 215,455 sequences was generated to assess the final performance and generalisation of the trained model. The test set used in this study is entirely independent of the train and validation datasets to ensure the reliability and generic aspect of the proposed framework.

Table 1 provides the classes distribution in each dataset. This table gives an overview of the distribution of the two different sequences classes (labeled as legitimate and malicious), present in the Training, Validation, and Test datasets. This information is critical for assessing the balance of the two classes within the datasets, which can impact the model's performance and lead to any required adjustments during the training and evaluation phases.

**Table 1.** Data distribution of the datasets used to evaluate the models. Dataset 1 is generated on a unique local network. Dataset 2 corresponds to the same scenarios, but where the SIP client and SIP server IP addresses are different from those used in the training phase.

|  | Dataset 1 (Training) | | | Dataset 2 (Inference) |
| --- | --- | --- | --- | --- |
|  | Train | Validation | Test | Test |
| Legitimate sequence | 678,343 | 44,844 | 124,019 | 124,019 |
| Malicious sequence | 457,512 | 32,827 | 91,436 | 91,436 |
| Total | 1,135,855 | 77,671 | 215,455 | 215,455 |

### 4.3   Features Selection

Table 2 presents the different main SIP fields, their corresponding data types in the dataset and the selected SIP fields as a features to be used by the model.

- Time (Temps): The reception time of the INVITE message is an essential feature for identifying the reception time of the sequence associated with potential DDoS attacks. By analyzing the reception times of INVITE messages, security analysts can identify patterns that may indicate a DDoS attack is in progress. For example, if a large number of INVITE messages are received within a short period of time, this may be indicative of a flooding attack.
- Source (IP address): This feature indicates the IP address of the client from the sent INVITE message. The source IP address is essential in pinpointing the origin of the message and help to identify suspicious or malicious sources of the DDoS attacks.
- From (SIP URI of the client): The "From" SIP-field includes the SIP Uniform Resource Identifier (URI) of the client, to identify the sender of a SIP request. This information helps in characterizing the clients involved in the communication and allow to distinguish legitimate clients from malicious attackers.
- Call-ID (URI of the call): The Call-ID field contains the URI of the session, providing a unique identifier for each session. This data is used for detecting patterns and correlations between different INVITE messages associated with the same session.
- Contact: This field identifies a SIP URI to which a SIP response can be sent. The SIP contact information can help us to know the client identity.

By selecting these fields as features for the training data, the models can learn and recognize patterns that reflect SIP-DDoS attacks. The combination of timing reception and the selected features (selected SIP fields) enhances the models' ability and effectiveness to differentiate between legitimate and malicious INVITE SIP traffic. The rationale behind excluding certain features lies in their lack of relevance for the considered scenarios. We carefully selected features to streamline the training phase, mitigates the risk of over-fitting and allow the models to focus on the most informative attributes.

**Table 2.** Dataset's column and selected features.

| SIP field | Data type | Selected features |
|---|---|---|
| Seq | object | ✗ |
| Temps | object | ✓ |
| Source | object | ✓ |
| MSG | object | ✗ |
| Via | object | ✗ |
| From | object | ✓ |
| To | object | ✗ |
| Call-ID | object | ✓ |
| CSeq | object | ✗ |
| Contact | object | ✓ |
| Max-Forwards | int64 | ✗ |
| Subject | object | ✗ |
| Content-Type | object | ✗ |
| Content-Length | int64 | ✗ |

### 4.4   Deep Learning Models

Within the IDS system, the implemented algorithms are: Recurrent Neural Network (RNN) [3], Long Short-Term Memory (LSTM) [4], and Gated Recurrent Unit (GRU) [2]. These algorithms are good candidates for sequence analysis and have proven to be effective in capturing temporal dependencies of data. By leveraging these algorithms, we can effectively analyze and detect malicious sequences in the SIP communication. The RNN architecture is designed to handle sequential data by processing information in a recurrent manner. LSTM, a variant of RNN, incorporates memory cells to capture long-term dependencies and prevent the vanishing gradient problem. Similarly, GRU also addresses the vanishing gradient problem while maintaining a simpler architecture compared to LSTM. By using RNNs, LSTM and GRU in the IDS-Sequence system, we can leverage their temporal modeling capabilities to capture the temporal properties of SIP data traffic. The pre-training span 10 epochs and used a batch size of 250. We implemented our approach using the PyTorch libraries and as our working environment we used Google Colab with "T4 GPU".

### 4.5   Evaluation Metrics

To evaluate the proposed IDS, the following metrics are used: accuracy, precision, recall, and F1-score. These metrics are used for assessing the performance of our detection model for classifying malicious and legitimate SIP-INVITE traffic. In addition to the above metrics and for the sequences based approach, the Intersection over Union metric (IoU) and detection time are also used for the sequences classification. IoU is a typical metric used for tasks related to object detection and segmentation. It measures the overlap between the predicted region and the ground truth region of an object. In the context of sequence classification, IoU can be utilized to determine the similarity between predicted and expected classifications.

Detection time is a critical metric used for measuring the time cost and efficiency of the proposed model for detecting malicious SIP-INVITE traffic. It assesses the delay between the time of the received sequence and the time the sequence is correctly classified as malicious or legitimate.

By exploiting these metrics, we provide an extensive and comprehensive performance evaluation for detecting and classifying the malicious and legitimate SIP traffic, with effectiveness.

$$\text{Accuracy} = \frac{\text{TP} + \text{TN}}{\text{TP} + \text{TN} + \text{FP} + \text{FN}} \tag{2}$$

$$\text{Precision} = \frac{\text{TP}}{(\text{TP} + \text{FP})} \tag{3}$$

$$\text{Recall} = \frac{\text{TP}}{(\text{TP} + \text{FN})} \tag{4}$$

$$\text{F1-score} = 2 \times \frac{\text{Precision} \times \text{Recall}}{\text{Precision} + \text{Recall}} \tag{5}$$

$$\text{IoU} = \frac{\text{TP}}{\text{TP} + \text{FP} + \text{FN}} \tag{6}$$

where True Positives (TP) correspond to the number of instances detected as true malicious traffic. True Negatives (TN) is the number of instances detected as false malicious traffic. False Positives (FP) is the number of instances the legitimate traffic is detected as a malicious traffic. And, False Negatives (FN) correspond to the number of instances the malicious traffic is detected as a legitimate traffic.

## 5   Evaluation

The results in terms of classification, based on the performance metrics, give a high rate for RNN, LSTM, and GRU for each dataset 1 and 2 (Accuracy = 99.9%, Precision = 99.9%, Recall = 99.9%, F1-score = 99.9% and IoU = 99.9%). Also, the performance gap between RNN, LSTM and GRU models is significantly low, with standard deviation of ±0.02. Given this, we propose to focus the analysis based on the comparison of the other metrics: FP, FN and "misclassifed sequence" and "inference time".

The Table 3 presents the results obtained from the RNN, LSTM and GRU models for the two distinct datasets. The evaluation is performed based on misclassified sequences, inference metrics, False Negatives (FN), and False Positives

(FP). As shown, GRU outperforms LSTM and RNN models in terms of misclassified sequences. In the dataset 1, the GRU model misclassifies only 63 sequences, demonstrating its superiority over the RNN and the LSTM models, which reach respectively 102 and 109 sequences. For the dataset 2, the GRU model also outperforms LSTM and GRU, with 48 misclassified sequences, while the LSTM misclassifies 85 sequences, and RNN misclassifies 103 sequences.

**Table 3.** Malicious sequence detection results on the two datasets.

| Dataset 1 | | | |
|---|---|---|---|
| | **RNN** | **LSTM** | **GRU** |
| Misclassified sequences | 102 | 109 | 63 |
| Inference Time (10 messages) | 700.00 µs | 749.00 µs | 763.00 µs |
| FP | 0.05 % | 0.1 % | 0.05 % |
| FN | 0.04 % | 0.01 % | 0.01 % |
| Dataset 2 | | | |
| | **RNN** | **LSTM** | **GRU** |
| Misclassified sequences | 103 | 85 | 48 |
| Inference Time (10 messages) | 695.00 µs | 753.00 µs | 757.00 µs |
| FP | 0.05 % | 0.06 % | 0.03 % |
| FN | 0.04 % | 0.01 % | 0.01 % |

From the inference metrics, the RNN models showcase better inference times. For dataset 1, RNN processes 10 INVITE messages within 700 µs, whereas the GRU model is slightly higher, with a processing time of 763 µs. For dataset 2, RNN shows a lower inference time of 695 µs, while GRU gives 757 µs for the inference time. Regarding the False Positives (FP) parameter, GRU also outperforms RNN and LSTM, showcasing its effectiveness in minimizing FP. For dataset 1, GRU achieves an height performance results, with a False Positive rate of 0.05%, equivalent to the RNN. For dataset 2, GRU maintains its superiority, achieving a False Positive rate of 0.03%. In contrast, RNN reports a slightly higher False Positive rate of 0.05%, while LSTM shows the highest False Positive rate of 0.06%. Regarding the False Negatives (FN) metric, GRU emerges as the front-runner, showcasing its capability to minimize missed classification. For both datasets 1 and 2, GRU achieves a significant low False Negative rate of 0.01%, outperforming RNN by 0.04% while matching the performance of LSTM.

Table 4 provides a detailed and comprehensive illustration regarding the distribution of misclassified sequences within the dataset. It highlights specific sequence patterns that consistently pose challenges for the classification models. These specifics sequences patterns are the most frequently misclassified patterns compared to the others and need further investigations.

In addition, it shows the frequency of misclassifications for GRU, LSTM and RNN. "T-1" and "T-2" in the table indicate their performance for Test datasets 1 and 2. Particularly the sequences (1, 1, 0, 0, 0, 0, 0, 0, 0, 0) and (0, 0, 0, 0, 0, 0, 0, 0, 1, 1) are of particular interest. Indeed, these sequences are somehow challenging for GRU, LSTM and RNN, as they are frequently misclassified. Furthermore, the sequence (0, 0, 0, 0, 0, 0, 0, 0, 1, 0) appears to be challenging for both LSTM and GRU, as they are also misclassified.

**Table 4.** Results for the frequency of pattern classification errors in dataset 1 (T-1) and dataset 2 (T-2).

| Sequence pattern | Train | Test | LSTM | | GRU | | RNN | |
|---|---|---|---|---|---|---|---|---|
| | | | T-1 | T-2 | T-1 | T-2 | T-1 | T-2 |
| (1, 0, 0, 0, 0, 0, 0, 0, 0, 0) | 500 | 95 | 0 | 7 | 0 | 0 | 21 | 29 |
| (1, 1, 0, 0, 0, 0, 0, 0, 0, 0) | 529 | 93 | 31 | 26 | 11 | 8 | 11 | 12 |
| (1, 1, 1, 0, 0, 0, 0, 0, 0, 0) | 628 | 110 | 9 | 6 | 0 | 2 | 0 | 0 |
| (1, 1, 1, 1, 0, 0, 0, 0, 0, 0) | 1,080 | 200 | 8 | 3 | 0 | 0 | 0 | 0 |
| (0, 0, 0, 0, 0, 0, 0, 0, 1, 0) | 31 | 7 | 5 | 5 | 7 | 6 | 0 | 0 |
| (0, 0, 0, 0, 0, 0, 0, 0, 1, 1) | 553 | 100 | 9 | 9 | 18 | 12 | 20 | 19 |
| | | | 62 | 56 | 36 | 28 | 52 | 54 |

In conclusion, the performance results highlight that GRU stands out as the most effective and efficient candidate for our classification and use case. Although the GRU model shows the lowest inference times (763 μs for Test dataset 1, 757 μs for Test dataset 2), the gap is not significant compared to the other models (LSTM and RNN) and does not impact the IoT-SIP sessions (few ms to few sec).

# 6   Conclusion

We have proposed a framework for detecting and mitigating SIP-INVITE flooding attacks in SIP based IoT infrastructure. We also demonstrate the effectiveness and robustness of our proposed approach through extensive and comprehensive experiments and evaluations under various simulated scenarios. The framework comprises a dataset generated exploiting SIPp and Mr.SIP tools, containing legitimate and malicious INVITE message simulations under various scenarios with different message rates. Scenarios range from 10 messages/s to 1000 messages/s, allowing comprehensive evaluation of the proposed framework. The core component of the proposed framework is a contribution for a GRU-based IDS designed to analyze incoming SIP traffic in real time. GRU constitutes a good candidate for capturing temporal patterns and characteristics in sequential data, while identifying INVITE flooding patterns. By leveraging our datasets for training, the GRU-based IDS is able to distinguish in real time and with high accuracy legitimate and malicious SIP-INVITE messages. The results confirm that the proposed approach and framework can be effectively deployed in SIP based IDS for IoT environments.

For short-term perspectives, we envision two directions: the first one is to investigate the impact of different step values on detection metrics, e.g., performance comparison for step value equal 5 and 10. The second direction is to deploy our proposed solution in real IoT based IDS platform and test-bed.

**Acknowledgment.** This work has been carried in the context of the project Beyond5G, funded by the French government as part of the economic recovery plan, namely "France Relance" and the investments for the future program.

# References

1. Alvares, C., Dinesh, D., Alvi, S., Gautam, T., Hasib, M., Raza, A.: Dataset of attacks on a live enterprise voip network for machine learning based intrusion detection and prevention systems. Comput. Netw. **197**, 108283 (2021)
2. Chung, J., Gulcehre, C., Cho, K., Bengio, Y.: Empirical evaluation of gated recurrent neural networks on sequence modeling. arXiv preprint arXiv:1412.3555 (2014)
3. Elman, J.L.: Finding structure in time. Cogn. Sci. **14**(2), 179–211 (1990)
4. Graves, A., Graves, A.: Long short-term memory. Supervised sequence labelling with recurrent neural networks, pp. 37–45 (2012)
5. Hussain, I., Djahel, S., Zhang, Z., Naït-Abdesselam, F.: A comprehensive study of flooding attack consequences and countermeasures in session initiation protocol (SIP). Secur. Commun. Netw. **8**(18), 4436–4451 (2015)
6. Inayat, U., Zia, M.F., Mahmood, S., Khalid, H.M., Benbouzid, M.: Learning-based methods for cyber attacks detection in IoT systems: a survey on methods, analysis, and future prospects. Electronics **11**(9), 1502 (2022)
7. Khalil, H., Elgazzar, K.: Leveraging blockchain for device registration and authentication in tSIP-based phone-of-things (PoT) systems. In: 2023 International Wireless Communications and Mobile Computing (IWCMC), pp. 1605–1612. IEEE (2023)
8. Kumari, P., Jain, A.K.: A comprehensive study of DDoS attacks over IoT network and their countermeasures. Comput. Secur. 103096 (2023)
9. Mahajan, N., Chauhan, A., Kumar, H., Kaushal, S., Sangaiah, A.K.: A deep learning approach to detection and mitigation of distributed denial of service attacks in high availability intelligent transport systems. Mob. Netw. Appl. **27**(4), 1423–1443 (2022)
10. Meddahi, A., Drira, H., Meddahi, A.: SIP-GAN: generative adversarial networks for sip traffic generation. In: 2021 International Symposium on Networks, Computers and Communications (ISNCC), pp. 1–6. IEEE (2021)
11. Meshram, C., Lee, C.C., Bahkali, I., Imoize, A.L.: An efficient fractional Chebyshev chaotic map-based three-factor session initiation protocol for the human-centered IoT architecture. Mathematics **11**(9), 2085 (2023)
12. Mittal, M., Kumar, K., Behal, S.: Deep learning approaches for detecting DDoS attacks: a systematic review. Soft Comput. 1–37 (2022)
13. Nassar, M., State, R., Festor, O.: Labeled VoIP data-set for intrusion detection evaluation. In: Aagesen, F.A., Knapskog, S.J. (eds.) EUNICE 2010. LNCS, vol. 6164, pp. 97–106. Springer, Heidelberg (2010). https://doi.org/10.1007/978-3-642-13971-0_10

14. Nazih, W., Hifny, Y., Elkilani, W.S., Dhahri, H., Abdelkader, T.: Countering DDoS attacks in sip based VoIP networks using recurrent neural networks. Sensors **20**(20), 5875 (2020)

15. Omolara, A.E., Alabdulatif, A., Abiodun, O.I., Alawida, M., Alabdulatif, A., Arshad, H., et al.: The internet of things security: a survey encompassing unexplored areas and new insights. Comput. Secur. **112**, 102494 (2022)

16. Pereira, D., Oliveira, R.: Detection of abnormal sip signaling patterns: a deep learning comparison. Computers **11**(2), 27 (2022)

17. Pereira, D., Oliveira, R., Kim, H.S.: Classification of abnormal signaling sip dialogs through deep learning. IEEE Access **9**, 165557–165567 (2021)

18. Rosenberg, J., et al.: SIP: session initiation protocol. Technical report (2002)

19. SIPp: Sipp. https://sipp.sourceforge.net/

20. Stanek, J., Kencl, L.: SIPp-DD: sip DDOS flood-attack simulation tool. In: 2011 Proceedings of 20th International Conference on Computer Communications and Networks (ICCCN), pp. 1–7. IEEE (2011)

21. Tas, I.M., Unsalver, B.G., Baktir, S.: A novel sip based distributed reflection denial-of-service attack and an effective defense mechanism. IEEE Access **8**, 112574–112584 (2020)

22. Yang, I.F., Lin, Y.C., Yang, S.R., Lin, P.: The implementation of a SIP-based service platform for 5G IoT applications. In: 2021 IEEE 93rd Vehicular Technology Conference (VTC2021-Spring), pp. 1–6. IEEE (2021)

23. Yang, S.R., Lin, Y.C., Lin, P., Fang, Y.: AioTtalk: a sip-based service platform for heterogeneous artificial intelligence of things applications. IEEE Internet Things J. (2023)

# Deep Reinforcement Learning
# for Multiobjective Scheduling in Industry
# 5.0 Reconfigurable Manufacturing Systems

Madani Bezoui[1] ⓘ, Abdelfatah Kermali[1] ⓘ, Ahcene Bounceur[2(✉)] ⓘ,
Saeed Mian Qaisar[3,4] ⓘ, and Abdulaziz Turki Almaktoom[4] ⓘ

[1] CESI LINEACT, UR 7527, Nice, France
{mbezoui,akermali}@cesi.fr
[2] KFUPM, ICS Department, Dhahran, Saudi Arabia
Ahcene.Bounceur@kfupm.edu.sa
[3] CESI LINEACT, UR 7527, Lyon, France
smianqaisar@cesi.fr
[4] Electrical and Computer Engineering Department, Effat University, Jeddah 22332,
Kingdom of Saudi Arabia

**Abstract.** In modern-day manufacturing, it is imperative to react
promptly to altering market requirements. Reconfigurable Manufacturing Systems (RMS) are a significant leap forward in achieving this criteria
as they offer a flexible and affordable structure to comply with evolving production necessities. The ever-changing nature of RMS demands a
sturdy induction of learning algorithms to persistently improve system
configurations and scheduling. This study suggests that using Reinforcement Learning (RL), specifically, the Double Deep Q-Network (DDQN)
algorithm, is a feasible way to navigate the intricate, multi-objective
optimization landscape of RMS. Key points to consider regarding this
study include cutting down tardiness costs, ensuring sustainability by
reducing wasted liquid and gas emissions during production, optimizing
makespan, and improving ergonomics by reducing operator intervention
during system reconfiguration. Our proposal consists of two layers. Initially, we suggest a hierarchical and modular architecture for RMS which
includes a multi-agent environment at the reconfigurable machine tool
level, which improves agent interaction for optimal global results. Secondly, we incorporate DDQN to navigate the multi-objective space in a
clever manner, resulting in more efficient and ergonomic reconfiguration
and scheduling. The findings indicate that employing RL can help solve
intricate optimization issues that come with contemporary manufacturing paradigms, clearing the path for Industry 5.0.

**Keywords:** Reconfigurable Manufacturing Systems · Sustainability ·
Deep Reinforcement Learning · Multiobjective Scheduling · Industry
5.0

The authors are thankful to the CESI LINEACT, Effat University, and King Fahd
University of Petroleum and Minerals for the technical support. They are also thankful
to the Effat University for financially supporting this project under the grant number
(UC No. 9/12June2023/7.1-21(4)11).

É. Renault et al. (Eds.): MLN 2023, LNCS 14525, pp. 90–107, 2024.
https://doi.org/10.1007/978-3-031-59933-0_7

# 1 Introduction

In an era of rapid technological evolution and fluctuating market dynamics, industries, particularly the manufacturing sector, are in a constant search for innovative solutions to meet the escalating demand for product customisation and to skilfully navigate the unpredictability of market trends [10]. In the face of these challenges, Reconfigurable Manufacturing Systems (RMS), a concept introduced by Koren et al. [6], has emerged as a seminal solution. RMS represents a pivotal evolution in manufacturing paradigms, seeking to merge harmoniously the high-throughput features of Dedicated Manufacturing Lines (DML) - characterised by the use of Dedicated Machines (DM) - with the adaptability and responsiveness of Flexible Manufacturing Systems (FMS), characterised by the use of Computer Numerical Control (CNC) machines. This transformation from DML to FMS and finally to RMS represents a profound shift in manufacturing philosophy, where adaptability, modularity and rapid reconfiguration are essential to meet the dynamic demands of today's industrial landscapes [6].

The core challenges associated with RMS are orchestrated around three key stages: design, implementation and optimisation. The design stage provides the fundamental blueprint that defines the architecture of the system, which in turn significantly influences the subsequent implementation and optimisation stages. Crucial decisions regarding system components and their potential reconfigurations are made during this phase [2]. The design phase is followed by the implementation phase, which translates theoretical designs into tangible, operational systems. This transition is fraught with intricacies, encapsulating the integration of diverse system components, ensuring operational coherence, and accommodating practical considerations [7]. Ultimately, the optimisation stage continually refines system operations to increase efficiency and productivity while reducing costs and waste. This stage employs a range of optimisation tools and strategies such as scheduling, machine allocation and production planning, and also requires the consideration of real-time data and prospective reconfigurations [3].

In recent years, the infusion of Artificial Intelligence (AI) and learning methods, particularly Reinforcement Learning (RL), has gained considerable traction in solving complex puzzles in various domains. These technological advancements promise to improve decision making, system adaptation and optimisation, thus addressing some of the critical challenges associated with RMS implementation.

# 2 Literature Review

The advent of Reconfigurable Manufacturing Systems (RMS) marked a significant milestone towards addressing the dynamic demands of modern manufacturing landscapes. The application of various optimization techniques within the RMS domain has garnered substantial attention, owing to the potential to address complex scheduling and reconfiguration problems.

Several studies have underscored the application of mathematical programming approaches like mixed-integer linear programming (MILP) in optimizing scheduling and reconfiguration in RMS settings. For instance, Aljuneidi and Bulgak [1] proposed a MILP model for designing reconfigurable cellular manufacturing systems with hybrid manufacturing-remanufacturing capabilities. The model incorporated constraints related to machine capacities, inventory levels, and processing requirements. Khezri et al. [4] developed a multi-objective MILP model to generate sustainable process plans in a reconfigurable context. The model incorporated objectives related to minimizing production cost, time as well as environmental impacts measured through liquid waste generation and greenhouse gas emissions. The model was solved using an augmented epsilon-constraint method. In another study, Khezri et al. [5] proposed a bi-level decomposition approach involving two MINLP models to integrate diagnosability and sustainability considerations in RMS design. The upper-level model focused on preventive maintenance planning to minimize hazardous energy consumption, while the lower-level model concerned sustainable process plan generation by minimizing energy losses. Several studies have also adapted metaheuristic techniques like genetic algorithms, particle swarm optimization, ant colony optimization to effectively solve optimization problems related to scheduling and reconfiguration in RMS settings under complexity constraints. Musharavati and Hamouda [9] developed a simulated annealing algorithm for optimizing process plans in RMS environments. The proposed approach outperformed a standard SA implementation in the comparative analysis. Mohapatra et al. [8] employed an adapted NSGA-II algorithm for integrating process planning and scheduling decisions in an RMS through adaptive setup planning. The objectives of minimizing completion time and cost were simultaneously optimized through the NSGA-II implementation.

The integration of artificial intelligence techniques like reinforcement learning, neural networks and fuzzy logic has also garnered attention for automating and enhancing RMS scheduling and reconfiguration capabilities. For instance, Yang and Xu [12] proposed a multi-agent deep reinforcement learning architecture for optimized scheduling and reconfiguration in smart RMS environments. The model design demonstrated the versatility of deep reinforcement learning for automated decision-making in complex RMS settings. Tang and Salonitis [11] explored a deep reinforcement learning-based scheduling policy tailored for reconfigurable manufacturing systems, elucidating the role of deep reinforcement learning in devising adaptive scheduling policies. Zhou et al. [14] presented a dynamic scheduling approach based on deep reinforcement learning to address scheduling challenges in smart manufacturing contexts enabled by RMS.

In summary, mathematical programming, metaheuristics and artificial intelligence have been widely leveraged to tackle scheduling and reconfiguration challenges in RMS environments, with the techniques offering complementary strengths. While mathematical models support optimization under precisely defined constraints, metaheuristics and AI facilitate solving highly complex problems through exploration of large search spaces. Further hybridization of these techniques can pave the way for more holistic and powerful RMS optimization capabilities.

# 3    Problem Description

The main focus of this work is to explore the complexity of scheduling and reconfiguration in the field of reconfigurable manufacturing systems, especially in the emerging framework of Industry 5.0. The problem is multi-objective in nature, involving the simultaneous optimisation of objectives such as minimising tardiness costs, reducing reconfiguration operator interventions, optimising makespan, and improving system ergonomics. This section unfolds the core aspects of the proposed modelling of the RMS architecture, the underlying multi-objective problem and the contextual relevance of Industry 5.0, providing a solid foundation for the proposed methodology and its subsequent implementation and experimentation.

## 3.1    Reconfigurable Manufacturing System (RMS)

The key components of RMS include reconfigurable machine tools (RMTs), modular equipment and intelligent control systems. These elements work in a complementary way to facilitate rapid reconfiguration, rescheduling and adaptation to new manufacturing tasks. The modular design of RMT enables a wide range of configurations to meet different product specifications and production volumes. Meanwhile, the intelligent control systems use advanced algorithms and real-time data to orchestrate the seamless transition between different configurations and schedules. The proposed hierarchical environment for RMS architecture consists of a top-level RMS environment that includes several lower-level RMT sub-environments. Each of these RMT sub-environments is as a multi-agent environment (Fig. 1).

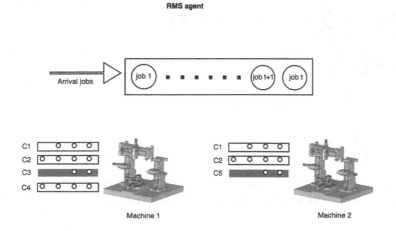

**Fig. 1.** RMS Modelling with RMT machines with differents configurations.

## 3.2  Reconfigurable Machine Tool (RMT)

Reconfigurable Machine Tool (RMT) accepts multiple configurations, each configuration is represented by a virtual buffer in this model. A configuration refers to a specific arrangement of the machine's modules to obtain certain capabilities. RMT is controlled by two cooperative but independent agents inspired from the work of [12]:

✓ The Reconfiguration Agent (RCA) handles dynamic Reconfiguration of the machine's modules among a predefined set of configurations to adapt its capabilities.
✓ The Scheduling Agent (SA) is responsible for optimizing the scheduling and sequencing of jobs on the machine in its current configuration.

By jointly learning coordinated policies, the RCA and SA agents can intelligently RCAigure the machine's modules and adjust the production SAule to optimize objectives like minimizing job tardiness as conditions change.

## 3.3  Reconfiguration Agent RCA

The objective is to select reconfiguration actions that minimize any rise in total tardiness costs of jobs during the Reconfiguration interval. Let $t_S$ and $t'_S$ be the start and end times of a reconfiguration.

The reward $R_t$ for a reconfiguration is:

$$R^{RCA} = -\frac{1}{t_{S'} - t_S}(TC_{WJ} + TC_{CJ}) \tag{1}$$

$$TC_{WJ} = \sum_{k=1}^{X}\sum_{j=1}^{n_k} \alpha_j z_{j_{S'}} \left[t_{S'} - \max(t_S, d_j)\right]$$

$$z_{j_{S'}} = \begin{cases} 1, \text{ if } d_j < t_{S'} \\ 0, \text{ else} \end{cases}$$

$$TC_{CJ} = \sum_{j=1}^{n_{FNS}} \alpha_j z_{jC} \left[C_j - \max(t_S, d_j)\right]$$

$$z_{jC} = \begin{cases} 1, \text{ if } d_j < C_j \\ 0, \text{ else.} \end{cases}$$

Where $TC_{WJ}$ and $TC_{CJ}$ are the newly added tardiness costs of waiting and completed jobs. This rewards actions that keep the increase in total tardiness low.

**Reconfiguration Trigger:** Reconfiguration is triggered when:

✓ The current buffer is empty
✓ Overdue jobs exist and the current buffer has relatively low average tardiness
✓ The number of finished jobs exceeds a threshold and average tardiness is low

This balances reconfiguration and scheduling optimization.

**Action Space:** Four actions are defined for selecting a new configuration $k$ from $1, 2, ..., K$:

✓ **Action 1** ($act1$): Choose $k$ with maximum total tardiness cost $\sum_j \beta_j$ of its jobs
✓ **Action 2** ($act2$): Choose $k$ with maximum average tardiness cost $\frac{1}{n_k} \sum_j \beta_j$
✓ **Action 3** ($act3$): Choose $k$ with maximum number of jobs $n_k$
✓ **Action 4** ($act4$): Choose $k$ with minimum average safe time $\frac{1}{n_k} \sum_j ST_j$

If selected $k$ matches current, no reconfiguration is done.

**State Features:** Four state features are used:

✓ $F_1$: Number of jobs in each configuration.
✓ $F_2$: Total tardiness cost in each configuration.
✓ $F_3$: Average tardiness cost in each configuration.
✓ $F_4$: Average safe time in each configuration.

The state space includes statistical measures on these features. States are normalized using min − max scaling.

## 3.4   Scheduling Agent SA

The goal of SA is selecting the next job from the current buffer (selected RMT configuration) to execute.

**Reward Function:** It is inversely related to the change in total tardiness per second:

$$R^{SA} = -\frac{1}{t_{s'} - t_s}(TC_{WF}) \tag{2}$$

Where $TC_{WF}$ is the newly added tardiness cost of waiting jobs in $[t_s, t_{s'}]$. This is calculated as:

$$TC_{WF} = \sum_{j=1}^{n} y_{js'} \beta_j [t_{s'} - \max(t_s, q_j)] \tag{3}$$

Where $y_{js'}$ indicates if job $j$ is overdue at $t_{s'}$. Rewarding lower incremental tardiness aims to have SA learn optimized job sequencing.

**Action Space:** The SA has five actions for selecting the next job $j$ from the current buffer:

✓ **Action 1** ($act1$): Choose job with maximum tardiness cost $\psi_j$.
✓ **Action 2** ($act2$): Choose job with maximum unit tardiness cost $\beta_j$.
✓ **Action 3** ($act3$): Choose job with minimum safe time $ST_j$.
✓ **Action 4** ($act4$): Choose job with nearest due date $q_j$.
✓ **Action 5** ($act5$): Choose job with minimum processing time $\sum_i t_{ij}$.

The diverse actions enable SA to learn nuanced scheduling policies.

**State Features:** Key state features for SA include:

✓ $F_1$: Number of waiting jobs.
✓ $F_2$: Current tardiness cost of each job.
✓ $F_3$: Unit tardiness cost of each job.
✓ $F_4$: Safe time of each job.
✓ $F_5$: Due date of each job
✓ $F_6$: Processing time of each job

Statistical measures on array features produce a 23-dim state space. States are normalized via $\min - \max$ scaling.

### 3.5   RMSA Agent

The RMSA supervises scheduling and reconfiguration decisions across multiple Reconfigurable Machine Tool (RMT) environments. This decentralised structure enables independent learning while reducing dimensionality. At each timestep, the RMSA assigns the next pending job to an RMT based on aggregated state data. Once initial jobs are assigned, the RMSA steps the RMTs to simulate system operation. This allows system oversight with machine autonomy.

**Multiobjective Scheduling and Reconfiguration Optimization:** The heart of the scheduling and reconfiguration problem in RMS is to determine the optimal sequence of operations on various machines, along with the optimal configuration path of machines and production lines to meet specified manufacturing objectives. The multiobjective nature of the problem requires a careful optimisation approach to find a Pareto-optimal solution that represents a balanced trade-off between the conflicting objectives. Each objective represents a critical dimension of operational performance and cost efficiency. For example, minimising the tardiness cost is critical to maintaining delivery scheduling and customer satisfaction, while reducing number of reconfiguration operator interventions is critical to improving system responsiveness to market changes. Optimising makespan has a direct impact on throughput and inventory levels, and improving system ergonomics is critical to ensuring operator safety and job satisfaction. Exploring these objectives, both individually and collectively, reveals a complex optimisation landscape with countless local optima and intricate constraints.

**Industry 5.0 Context:** The advent of Industry 5.0 heralds a human-centred approach to manufacturing, with an emphasis on fostering collaboration between people and intelligent systems. Unlike previous industrial revolutions, Industry 5.0 seeks to harmonise the strengths of human creativity and machine precision. The problem described in this paper is directly aligned with the ethos of Industry 5.0, as it seeks to optimise the operational efficiency of RMS while promoting ergonomic considerations for human operators. The envisaged solution aims to harness the capabilities of intelligent algorithms to not only enhance manufacturing performance, but also to create a conducive and ergonomic working environment. The synergy between man and machine is seen as a linchpin for achieving sustainable and inclusive growth in the manufacturing sector.

**Multiobjective Reward:** The RMSA agent employs a weighted multiobjective reward function to optimize multiple performance goals:

$$R^{\text{RMSA}} = w_1 F_t - w_2 F_s - w_3 F_o \qquad (4)$$

Where:

✓ $F_t$ = Average of RMTs agents' Tardiness cost reward
✓ $F_s$ = Sustainability objective
✓ $F_o$ = Operator intervention objective
✓ $w_1, w_2, w_3$ = Weight factors, which indicated the importance of each criterion. Usually, these weights are given by the Decision Maker.

The tardiness cost objective $F_t$ aims to minimize job delays. The sustainability objective $F_s$ encourages energy-efficient operation with lower emissions. The operator intervention objective $F_o$ penalizes excessive reconfigurations. By combining these objectives with tunable weights, the multiobjective reward allows the RMSA agent to learn policies that balance productivity, efficiency, human effort and other critical goals. The weights enable prioritizing the different objectives as needed for the manufacturing application.

Optimizing this composite reward function leads to Pareto optimal solutions that make trade-offs between the individual objective costs. This provides coordinated reconfiguration and scheduling control to maximize overall system performance across key dimensions.

**Tardiness Cost:** The RMSA tardiness cost reward is the average of the RMT tardiness cost rewards at each timestep. This aligns the global objective of maximizing productivity with the RMT rewards to minimize tardiness.

$$F_t = \frac{1}{N} \sum_{i=1}^{N} R_i^{RMT} \tag{5}$$

where:

$$R_i^{RMT} = \frac{R_i^{RCA} + R_i^{SA}}{2} \tag{6}$$

Where $N$ is the number of RMTs.

**Sustainability Objective:** The sustainability objective $F_s$ in the RMSA reward function aims to minimize the environmental impact of the manufacturing system operations. It accounts for both liquid hazardous waste (LHW) and greenhouse gas (GHG) emissions associated with each job and machine configuration inspired from the work of [4].

Where the sustainability cost $F_S$ is:

$$F_S = w_{\text{LHW}} \sum_{i=1}^{N} \sum_{j=1}^{n_i} \frac{(LHW_{ij} - LHW_{\text{min}})}{(LHW_{\text{max}} - LHW_{\text{min}})} + w_{\text{GHG}} \sum_{i=1}^{N} \sum_{j=1}^{n_i} \frac{(GHG_{ij} - GHG_{\text{min}})}{(GHG_{\text{max}} - GHG_{\text{min}})} \tag{7}$$

Where:

✓ $N$ is the number of RMT environments
✓ $n_i$ is the number of finished jobs in RMT $i$
✓ $LHW_{ij}$ and $GHG_{ij}$ are the LHW and GHG emissions for job $j$ in RMT $i$
✓ $LHW_{\text{min}}, LHW_{\text{max}}, GHG_{\text{min}}, GHG_{\text{max}}$ are normalization constants
✓ $w_{\text{LHW}}, w_{\text{GHG}}$ are objective weights

Minimizing this sustainability cost encourages configurations and job assignments that reduce hazardous wastes and emissions. The weights allow tuning the relative importance of LHW versus GHG reduction. Normalization accounts for variability in emission levels across jobs and machines.

**Operator Intervention Objective:** The operator intervention objective $F_o$ aims to minimize the total number of reconfigurations across all RMT environments. It penalizes frequent machine reconfiguration that requires human effort.

The operator intervention cost is:

$$F_o = \sum_{i=1}^{M} N_{r,i} \tag{8}$$

Where $M$ is the number of RMTs and $N_{r,i}$ is the number of reconfigurations performed in RMT $i$.

The total reconfigurations $N_r$ is calculated by summing the reconfiguration counts $N_{r,i}$ from each RMT environment $i$.

By penalizing the total reconfiguration count through this objective term, the RMSA agent is incentivized to learn policies that minimize operator effort required for machine reconfiguration across the manufacturing system. The weight $w_r$ controls the relative importance of this goal.

**Action Space:** The RMSA actions assign each pending job to an RMT selected based on aggregated states to optimize system performance.

There are 5 actions corresponding to RMT with:

✓ minimum configuration buffer length
✓ maximum configuration buffer safe time
✓ minimum configuration buffer total tardiness
✓ minimum configuration buffer average tardiness

RMT already running the configuration, or minimum length.

By considering factors like queue occupancy, safe time, and tardiness, the RMSA can improve system scheduling.

**State Features:** The RMSA state summarizes high-level features across RMTs, including normalized:

✓ Queue lengths per RMT
✓ Total tardiness per RMT
✓ Average tardiness per RMT
✓ Job safe times per RMT.

It tracks RMT metrics like queue occupancy and scheduling to inform system job assignments. The compact representation allows assessing global status for joint reconfiguration and scheduling.

# 4    Implementation and Experimentation

## 4.1    Implementation of DDQN

The DDQN algorithm, an extension of the conventional Deep Q-Networks (DQN) algorithm, is employed to tackle the multiobjective optimization problem at hand. DDQN mitigates the overestimation bias of Q-values, a known issue in standard Q-learning and DQN, by decoupling the selection and evaluation of actions.

The core components of the DDQN algorithm include:

✓ **State Representation**: The state of the system is represented using a set of features that encapsulate the current configuration and status of the RMS.
✓ **Action Space**: The action space comprises all possible reconfiguration and scheduling actions that can be executed at any given time.
✓ **Reward Function**: The reward function is designed to reflect the objectives of the optimization problem, such as minimizing tardiness costs, optimizing makespan, and enhancing system ergonomics.
✓ **Q-Network and Target Q-Network**: Two separate neural networks are employed to approximate the Q-values of state-action pairs, aiding in reducing overestimation bias.

The DDQN algorithm in this work is implemented in Python 3.10 version using GYMNASIUM API and Ray library tailored for deep reinforcement learning applications, in a machine with 1 GPU and 24 CPU and 32 GO of RAM. The implementation follows a systematic procedure:

1. **Initialization**: Initializing the Q-network and Target Q-network with random weights, and setting initial values for the learning rate $= 4e^{-4}$ and exploration rate $= 8e^{-4}$, which are the defaults used values for *Ray Library*.
2. **Training**: Training the DDQN algorithm using a replay buffer (with length of 5000) to store and sample experience tuples, and updating the Q-network weights using mini-batch gradient descent.
3. **Evaluation**: Evaluating the trained DDQN model on a set of test instances, generated like shown in Algorithm 2 to assess its performance in solving the scheduling and reconfiguration problem.

The algorithm begins by initialising the RMS agent with random weights and proceeds to execute a series of RMS episodes. Within each episode, the RMS environment is reset, and for each RMT environment in the set, machine configurations and job distributions are sampled. The algorithm maintains states for both the RMS and RMT environments, updating them as actions are taken.

---

**Algorithm 1:** Hierarchical RL for RMS and RMT

---

**Input:** RMS environment $E_{\mathrm{RMS}}$, Set of RMT environments $\{E_{\mathrm{RMT}}\}$, Initial jobs Learning rates $\alpha_{\mathrm{RMSA}}, \alpha_{RCA}, \alpha_{SA}$

1 Initialize RMSA with random weights $\theta_{\mathrm{RMS}}$;
2 **for** *each RMS episode* **do**
3     Reset $E_{\mathrm{RMS}}$;
4     **for** *each $E_{RMT}$ in $\{E_{RMT}\}$* **do**
5         Sample machine configurations;
6         Sample new job distribution;
7         Initialize state $s_0$;
8         Reset buffers $B$, completed jobs $J$;

9     Aggregate info from $\{E_{\mathrm{RMT}}\}$ into $s_{0,\mathrm{RMS}}$;
10    **while** *$E_{RMS}$ not done* **do**
11        Observe $s_{t,\mathrm{RMS}}$;
12        Sample $a_{t,\mathrm{RMS}} \sim A_{\mathrm{RMS}}(s_{t,\mathrm{RMS}}; \theta_{\mathrm{RMS}})$;
13        **for** *each $E_{RMT}$ in $\{E_{RMT}\}$* **do**
14            **if** *RCA triggered reconfiguration* **then**
15                Calculate reconfiguration time $\Delta t$;
16                Update state $s_{t+\Delta t,\mathrm{RMT}}$;
17                Reset $E_{\mathrm{RMT}}$;
18                Sample new machine configuration;
19                Sample new job distribution;
20                Reset $B$, $J$;
21                Initialize state $s_{0,\mathrm{RMT}}$;
22                Update aggregated RMS state $s_{t+\Delta t,\mathrm{RMS}}$;

23            **while** *$E_{RMT}$ not done* **do**
24                Observe state $s_{t,\mathrm{RMT}}$;
25                Sample $a_{t,\mathrm{RCA}} \sim RCA(s_{t,\mathrm{RMT}}; \theta RCA)$;
26                Sample $a_{t,\mathrm{SA}} \sim SA(s_{t,\mathrm{RMT}}; \theta SA)$;
27                Execute $a_{t,\mathrm{RCA}}, a_{t,\mathrm{SA}}$ in $E_{\mathrm{RMT}}$;
28                Update $s_{t+1,\mathrm{RMT}}, B, J$;
29                Calculate rewards $r_{t,\mathrm{RCA}}, r_{t,\mathrm{SA}}$;
30                Aggregate RMT info into $s_{t+1,\mathrm{RMS}}$;

31        Store $(s_{t,\mathrm{RMS}}, a_{t,\mathrm{RMS}}, r_t, s_{t+1,\mathrm{RMS}})$;
32        Train RMSA on batch from replay buffer with learning rate $\alpha_{\mathrm{RMS}}$;
33        Update $\theta_{\mathrm{RMS}}$;

34 Train all RMT agents $\{RCA, SA\}$ on batches with learning rates $\alpha_{RCA}, \alpha_{SA}$;

---

The hierarchical structure becomes apparent as the algorithm handles reconfigurations initiated by the RCA agents and scheduling decisions made by the SA agents within the RMT environments. Information is aggregated between the RMS and RMT environments to enable holistic decision making. Throughout the process, the RL agents learn from their experience and adapt their policies. The algorithm demonstrates a multi-agent, hierarchical approach to resource management, which can be particularly useful in complex systems where coordination and decision making span multiple levels.

## 4.2  Experimental Instances Generation

In this algorithm, we first generate a set of jobs with random parameters including unit tardiness cost in this interval, due date, completion time, configuration choices, load handling weights, and greenhouse gas emissions. All these values are mentioned in Algorithm 2. We then create a Job object for each job instance and store it in the list of jobs. Finally, we provide a comment on how to represent the job's attributes, including its name, due date, and completion time.

---

**Algorithm 2:** Input Jobs Generation

---

**Data:** Number of Jobs $num\_of\_jobs$, Available Configurations $available\_configs$

**Result:** List of generated jobs $jobs$

35  **for** each $j$ in range $num\_of\_jobs$ **do**
36      Generate a unique job name $name$ based on $j$;
37      Generate unit tardiness cost $unit\_tard\_cost$ randomly between 2 and 10;
38      Generate due date $due\_date$ randomly between 10 and 50 after the arrival of the job;
39      Generate completion time $completion\_time$ randomly between 3 and 20;
40      Generate the number of configuration choices $num\_choices$ randomly between 1 and 2;
41      Generate $num\_choices$ random configuration choices $configs$ from $available\_configs$;
42      Generate quantity of liquid emitted by job of each configuration $configs\_lhw$;
43      Generate $num\_choices$ random configurations' greenhouse gas emissions of each configuration $configs\_ghg$;
44      Create a job object $job$ with attributes $name$, $unit\_tard\_cost$, $due\_date$, $completion\_time$, $configs$, $configs\_lhw$, $configs\_ghg$;
45      Append $job$ to the list of jobs $jobs$;

---

## 4.3  Training Methodology

In our exploration of training efficiency, we adopted a segmented approach, particularly focusing on the Reconfigurable Manufacturing Systems (RMS) and the Reconfigurable Manufacturing Tools (RMTs) within a hierarchical environment. Initially, we directed our attention towards individually training the multi-agent system associated with RMTs. A meticulous process of hyperparameter tuning was undertaken to find the optimal settings for the Learning Rate (LR) and the exploration factor $\varepsilon$. This tuning process employed a grid search technique, which is a systematic way of traversing through a manually specified subset of the hyperparameter space. The grid search mechanism allowed us to scrutinize various combinations of LR and $\varepsilon$ to pinpoint the configuration that yielded superior training outcomes.

The training curves for the Reconfiguration Agent (RCA) in Fig. 2a demonstrate its ability to learn effective policies for initiating reconfiguration actions based on the state of the manufacturing system. The smoothing trend of the reconfiguration time metric indicates the agent's proficiency in generalizing its experience to make timely and impactful reconfiguration decisions.

For the scheduling Agent (SA), Fig. 2b exhibits convergence in the tardiness cost. This highlights the agent's competence in developing robust job scheduling sequences that minimize delays and workflow duration even as manufacturing conditions evolve. The stability of the learning process enables the SA agent to consistently generate high-quality solutions, as noted by [13] in their work on deep reinforcement learning for job shop scheduling.

The tuning config was defined as:

```
config["search_space"] = {
"lr": tune.grid_search([1e-4, 2e-4, 3e-4, 4e-5, 5e-4, 6e-4]),
"epsilon": tune.grid_search([6e-4, 7e-4, 8e-4, 9e-4])
}
```

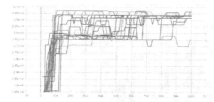

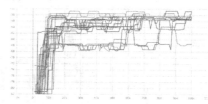

(a) Reconfiguration agent (RCA) training curves

(b) Scheduling agent (SA) training curves

**Fig. 2.** Training curves for RCA and SA

Upon obtaining the desired tuning, the training proceeded, during which the Q-Network and Target Network were evolved and refined. These networks embody the core learning mechanisms that guide the agents in making intelligent decisions. Once the training of the RMT multi-agent system reached a satisfactory level, the trained networks were preserved.

Transitioning to the broader hierarchical environment of RMS, we integrated the pre-trained networks from the RMT multi-agent system. This integration serves as a bedrock, providing a substantial head start for the agents operating within the RMS environment. By leveraging the insights and learned behaviors encapsulated in the pre-trained networks, the RMS hierarchical agent could navigate the environment with a higher degree of competency right from the outset. This strategy significantly expedited the training phase, fostering a more efficient learning trajectory and accelerating the attainment of desirable performance levels within the RMS hierarchical environment.

## 5   Results and Discussion

The results section encapsulates the insights obtained from the implementation and experimentation phase. Various performance metrics were analyzed to ascertain the effectiveness and efficiency of the proposed DDQN algorithm in managing multi-objective scheduling and reconfiguration challenges in reconfigurable manufacturing systems.

At the higher level, the manufacturing system agent (RMSA) in Fig. 3 demonstrates its capability to learn meta-policies that coordinate the lower-level RCA and SA agents to enhance overall system performance. The simultaneous optimization across all key metrics shows the agent's proficiency in balancing trade-offs through hierarchical reinforcement learning.

In summary, the empirical results provide quantitative evidence for the efficacy of the proposed hierarchical DDQN technique in enabling intelligent reconfiguration planning, scheduling, and coordination for next-generation reconfigurable manufacturing systems.

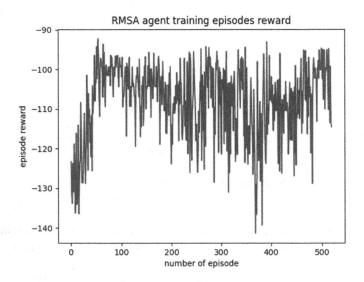

**Fig. 3.** Manufacturing system agent (RMSA) training curves

Additionally, the Pareto frontier in Fig. 4 obtained through the multi-objective DDQN approach exhibits a rich set of non-dominated solutions that make strategic compromises between the conflicting metrics. This highlights the adaptability of the intelligent agents in tailoring decisions to dynamic manufacturing conditions.

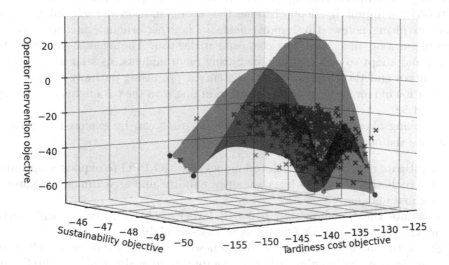

**Fig. 4.** Pareto frontier of non-dominated solutions

## 5.1 Discussion on Findings

The findings from the experimentation phase underscore the potential of deep reinforcement learning, particularly the DDQN algorithm, in revolutionizing resource management in reconfigurable manufacturing systems. The ability of the DDQN algorithm to dynamically adapt to changing manufacturing conditions, optimize resource allocation, and significantly improve operational efficiency elucidates its promise as a formidable tool for modern manufacturing environments.

Moreover, the hierarchical structure of the algorithm, as delineated in the implementation section, exhibits a nuanced approach to managing multi-level decision-making processes in complex manufacturing settings. This hierarchical approach enables a harmonized coordination between scheduling and reconfiguration decisions, thereby contributing to enhanced system performance. The success of the DDQN algorithm in both simulated and real-world environments substantiates the viability of leveraging advanced machine learning techniques to tackle intricate scheduling and reconfiguration challenges inherent in reconfigurable manufacturing systems. The positive outcomes from this project pave the way for further exploration and development of intelligent solutions for modern manufacturing challenges, aligning with the broader objectives of Industry 5.0 and smart manufacturing.

# 6    Conclusion and Future Work

The findings from this study underscore the potential of leveraging Hierarchical Deep Q-Networks (HDQNs) in addressing the intricate challenges inherent in reconfigurable manufacturing systems (RMS). The proposed framework, through a two-tiered approach, adeptly handles the reconfiguration and scheduling tasks, demonstrating noteworthy improvements in key performance metrics such as reconfiguration time, tardiness cost, and makespan. The agents' capability to learn and adapt to dynamic manufacturing environments, as seen in the training curves and the Pareto frontier, presents a promising pathway towards the realization of more agile and efficient RMS in line with the paradigms of Industry 4.0 and 5.0.

Looking ahead, there are several avenues that can be explored to further enhance the proposed framework:

✓ Investigate multi-objective RL techniques like MO-PPO to explicitly optimize for conflicting objectives like cost, sustainability, and ergonomics in a pareto-optimal manner.
✓ Evaluate the framework on real and larger problem instances with more machines, jobs, and configurations to assess scalability.
✓ Study the integration of the DRL framework with other Industry 4.0 technologies like digital twins, IoT, edge computing to enable real-time intelligent optimization.
✓ Analyze the reliability, robustness and worst-case performance of the DRL agents using formal verification methods.
✓ Evaluate alternate DRL algorithm architectures like attention-based transformers for RMS optimization. Assess interpretability.
✓ Benchmark against traditional optimization techniques like MILP and metaheuristics on extensive problem sets to gain deeper insights.

# References

1. Aljuneidi, T., Bulgak, A.: Designing a cellular manufacturing system featuring remanufacturing, recycling, and disposal options: a mathematical modeling approach. CIRP J. Manuf. Sci. Technol. **19**, 05 (2017)
2. Bilberg, A., Malik, R., Bøgh, K.: New model for development and manufacturing of tailored solutions in the industrial market. J. Manuf. Syst. **31**(3), 367–374 (2012)
3. Dashchenko, A.I.: Reconfigurable Manufacturing Systems and Transformable Factories. Springer, Heidelberg (2006)
4. Khezri, A., Benderbal, H.H., Benyoucef, L.: Towards a sustainable reconfigurable manufacturing system (SRMS): multi-objective based approaches for process plan generation problem. Int. J. Prod. Res. **59**(15), 4533–4558 (2021)
5. Khezri, A., Benderbal, H.H., Benyoucef, L., Dolgui, A.: Diagnosis on energy and sustainability of reconfigurable manufacturing system (RMS) design: a bi-level decomposition approach, September 2020
6. Koren, Y., Heisel, U., Jovane, F., et al.: Reconfigurable manufacturing systems. CIRP Ann. **48**(2), 527–540 (1999)

7. Li, Z., Li, L., Bilberg, A.: Design and implementation of a reconfigurable manufacturing system. Int. J. Adv. Manuf. Technol. **39**, 1181–1191 (2008)
8. Mohapatra, P., Benyoucef, L., Tiwari, M.: Realising process planning and scheduling integration through adaptive setup planning. Int. J. Prod. Res. **51**, 04 (2013)
9. Musharavati, F., Hamouda, A.M.S.: Simulated annealing with auxiliary knowledge for process planning optimization in reconfigurable manufacturing. Robot. Comput.-Integr. Manuf. **28**, 113–131 (2012)
10. Rajkumar, R., Ravi, G., Zalzala, A.: Recent advances in evolutionary and adaptable manufacturing systems. Int. J. Prod. Res. **48**(22), 6675–6696 (2010)
11. Tang, J., Haddad, Y., Salonitis, K.: Reconfigurable manufacturing system scheduling: a deep reinforcement learning approach. Procedia CIRP **107**, 1198–1203 (2022)
12. Yang, S., Xu, Z.: Intelligent scheduling and reconfiguration via deep reinforcement learning in smart manufacturing. Int. J. Prod. Res. **60**, 4936–4953 (2021)
13. Zhang, C., Song, W., Cao, Z., Zhang, J., Tan, P.S., Chi, X.: Learning to dispatch for job shop scheduling via deep reinforcement learning. In: Advances in Neural Information Processing Systems 33, pp. 1621–1632 (2020)
14. Zhou, L., Zhang, L., Horn, B.K.P.: Deep reinforcement learning-based dynamic scheduling in smart manufacturing. Procedia CIRP **93**, 383–388 (2020)

# Toward A Digital Twin IoT for the Validation of AI Algorithms in Smart-City Applications

Hamza Ngadi[1], Ahcene Bounceur[2(✉)], Madani Bezoui[3], Laaziz Lahlou[4],
Mohammad Hammoudeh[2], Kara Nadjia[4], and Laurent Nana[1]

[1] Lab-STICC, UBO, Brest, France
`Hamza.Ngadi@univ-brest.fr`
[2] Information and Computer Science Department, King Fahd University
of Petroleum and Minerals (KFUPM), Dhahran, Saudi Arabia
`{Ahcene.Bounceur,m.hammoudeh}@kfupm.edu.sa`
[3] CESI LINEACT, UR 7527, Nice, France
`mbezoui@cesi.fr`
[4] Ecole de Technologie Superieure, Montreal, Canada
`laaziz.lahlou@etsmtl.ca`

**Abstract.** The development of digital twins for road traffic has garnered significant attention within the scientific community, particularly in the realms of virtualization and the Internet of Things (IoT). The implementation of a digital twin for automobiles offers a virtual replica, capable of discerning the precise location, status, and real-time behavior of each vehicle present in the road traffic network. The primary objective of this endeavor is to create an advanced digital twin of cars that can seamlessly navigate through road traffic. To accomplish this, a meticulously selected technical approach involves employing a platform that simulates virtual sensor networks, accompanied by a purpose-built application that facilitates the access and dissemination of car-related data. Furthermore, this undertaking incorporates the utilization of an existing traffic simulator alongside a robust communication protocol to ensure seamless data transfer between the simulation environment and the sensors responsible for data collection from automobiles.

**Keywords:** Digital twin · Sumo simulator · Internet of Things · CupCarbon · MQTT communication protocol · Smart-city

## 1 Introduction

The development of a digital twin aims to duplicate a real object or environment taking into account its behavior over time, and then project it into a virtual environment for the purpose of monitoring or collecting data from the real environment. Data collection offers a means approved by scientific research in this field, to understand and analyze the object or environment targeted, in order to create predictions and scenarios of activities, whether undesirable or not from

E. Renault et al. (Eds.): MLN 2023, LNCS 14525, pp. 108–117, 2024.
https://doi.org/10.1007/978-3-031-59933-0_8

the real source environment [2]. The main motivations for creating a digital twin in a specific domain [8], can be driven by:

✓ Ensuring the reliability of behaviors and actions in an environment.
✓ Creating scenarios to guarantee stability in an environment.
✓ Guarantee the confidence of interaction with an object or an environment.
✓ Reducing and anticipating future failures of an object or environment.

From a technical point of view, a digital twin is a model that represents a description, a behavior and a state in time, updated with the shared data of an object, or an entity to be studied [7]. Several scales of creation of a digital twin are possible, these depend on the complexity of the objects or environment to simulate.

The scales can be varied between objects or equipment, complex systems, districts or cities up to the scale of a whole country, approached in the near future by the progress of scientific research [28].

To succeed in simulating the dynamics of road traffic that consists of at least one or more cars, the most common way in technical implementations is to equip one or more cars with sensors, thus, collecting their geolocation coordinates in real road traffic via these sensors, in order to use them to create or update the coordinates of the simulated cars in an application that implements a geographical map dedicated to visualization.

The technical implementation addressed in this work aims at using a Sumo road traffic simulator [3,22], to extract geographic data and send it via a communication protocol to an IoT simulator [23], which receives this data and updates the locations of the cars in the geographic area from which the cars are traveling. The presence of IoT sensors brings to life the digital twin of a city-related environment, providing a seamless and transparent means of exchanging data. Connected IoT objects offer the possibility for data to flow between real and virtual environments.

The collection and monitoring of data via connected sensors, give a global view of the interactions of an environment belonging to a city [13,23]. In this case, the construction of a twin with the behavior associated with each object of the targeted environment becomes possible, to simulate and study possible scenarios. The Sumo multimodal traffic simulator used in this work allows the creation and management of a large number of traffic networks, including the modeling and simulation of roads, vehicles and pedestrians. The data extracted from this tool belong to the simulations created from an OpenStreetMap geographical map, describing the behavior of cars in a road network.

OpenStreetMap (OSM) project [10,24] aims to build an online mapping database using open source geographic data. Common uses of this data are route planning, making available geographic data of real addresses and places, and accessing already collected data of a region, city or country in the form of a map, via map editors, libraries or APIs. The data extracted from a Sumo simulation is intended to be sent to the CupCarbon IoT simulator [5,18] for visualization of the cars in the IoT geographic map generated by this tool (Fig. 1).

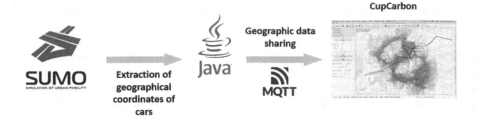

**Fig. 1.** Global process map of the developed application.

The following sections of this paper will discuss the creation of the application for data extraction, the implemented means of data sharing, and the receipt of data via the CupCarbon tool for visualization of the simulated cars, as well as the potential for improvement of the presented work.

## 2    Traffic Extractor Application

Internet of Things (IoT) simulation within digital twins enhances the efficiency and reliability of intelligent transport systems by enabling real-time data analysis and predictive modeling for optimized traffic management [12]. To establish a comprehensive repository of geographical data intended for utilization in the CupCarbon tool, aiming to construct a digital twin of road traffic relying on Sumo simulations, the imperative lies in crafting an intermediary application. This application shall seamlessly facilitate the exchange of data between the Sumo dataset and the CupCarbon IoT platform, ensuring a smooth and coherent flow of information between these two distinct entities.

Access to the simulation file plays a pivotal role in effectively leveraging the valuable data contained within the Sumo files. In order to achieve seamless data transfer from Sumo to the CupCarbon tool, a carefully considered communication protocol is adopted. This protocol serves as the conduit through which the data is efficiently transmitted towards the CupCarbon platform. To enhance user experience and foster intuitive interactions, a user-friendly graphical interface is meticulously designed and integrated into the application. This interface empowers users with the freedom to customize and fine-tune various simulation parameters, enabling them to execute scenarios according to their specific requirements. The thoughtful incorporation of this graphical interface not only simplifies user interactions but also streamlines the process of scenario simulation, contributing to an optimized and sophisticated workflow.

### 2.1    Access to the Scenario of A Sumo Simulation: Data Extraction

The Sumo simulator serves as a powerful tool capable of generating simulation files that encapsulate scenario-specific parameters for our chosen road traffic setup. The tool's comprehensive documentation offers access to these files through a variety of programming languages, such as Python, C++, and Java.

In the context of this work, Java has been selected as the language of choice for developing the application. To access the simulation files effectively, Sumo provides several libraries, and for this particular application, the Libsumo library [14,16,20] is employed, replacing the TraaS library [11,21], which relies on the TraCI (Traffic Control Interface) API [1,21] for accessing road traffic data from the simulation files. This ensures seamless and efficient retrieval of the required data for further processing within the application, thereby contributing to the successful realization of the project objectives.

The Libsumo library effectively fulfills the role previously performed by the TraCI API, serving as an intermediary interface between the Java application and the simulation. Within the developed Java application, it operates as a client, diligently querying the simulation to retrieve real-time information pertaining to the current states of the cars within the road environment specified by the generated Sumo file.

The Java application comprehensively captures the simulation by targeting the specific Sumo file, and in an iterative manner, it systematically traverses each round of the loop to extract pertinent data from one or more currently simulated cars. Subsequently, this extracted information is seamlessly transmitted to the CupCarbon tool, facilitating the visualization of the cars' movements and activities.

Upon completion of the scenario iterations, the Sumo simulation file is safely released, and the simulation itself is gracefully concluded. By effectively coordinating this process, the Java application successfully synchronizes data extraction and transmission, thereby optimizing the generation of the digital twin for road traffic in CupCarbon (Fig. 2).

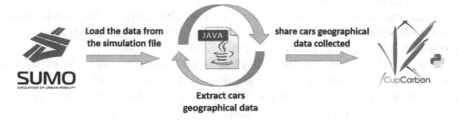

**Fig. 2.** The process of extracting geographical data from cars, from a Sumo simulation.

## 2.2   Data Sharing: MQTT Communication Protocol

In our technical implementation, the crucial aspect of data sharing is facilitated by the MQTT protocol [25,26]. This protocol is specifically designed for machine-to-machine communication, where relevant stakeholders subscribe to a centralized broker to facilitate a seamless and efficient exchange of data. The decision to adopt MQTT as the communication protocol is well-founded, owing to its user-friendly nature and streamlined implementation process.

One of the primary advantages of MQTT lies in its lightweight design, ensuring minimal energy consumption [15] and bandwidth usage. This inherent efficiency makes it an ideal and well-suited choice for IoT applications, such as our digital twin system for road traffic. The utilization of MQTT significantly contributes to the optimization of data transmission, promoting a highly responsive and agile ecosystem, ultimately enhancing the overall performance and effectiveness of the digital twin implementation.

Ensuring robust communication security is paramount for IoT in digital twin simulations, as it safeguards sensitive data and critical infrastructure, mitigates cyber threats, and ensures the integrity and reliability of the simulation environment [6]. In our project's architecture, the local broker selected for implementation is Mosquitto [4,9,19]. Mosquitto aligns with the inherent lightweight nature of the MQTT protocol, and it stands out for its effortless installation process and user-friendly operation. The advantage of Mosquitto lies in its ability to maintain the efficiency and responsiveness synonymous with MQTT, while simultaneously conserving system resources.

By incorporating Mosquitto as the local broker, we ensure seamless and efficient data communication between the various components of our digital twin system for road traffic. Its stability, ease of setup, and optimal utilization of machine resources make it an apt choice for facilitating reliable data exchange, thereby enhancing the overall performance and robustness of our IoT-based simulation.

A broker serves as an intermediary between clients utilizing protocols like MQTT, ensuring the reliable delivery and reception of messages, with adjustable quality of service in different shared topics. However, public and internet-accessible brokers may encounter challenges in message transmission and reception when facing saturation.

Apart from Mosquitto, there are several other alternatives that provide access to and utilization of the MQTT protocol. These widely used alternatives include: ActiveMQ [17], Apache Kafka [27], EMQX [4,9], VerneMQ [4,9], RabbitMQ [17], and HiveMQ [4,9]. Notably, HiveMQ is recognized for its stability and is often employed for online data sharing purposes. Each of these alternatives offers distinctive features and capabilities, allowing users to choose the most suitable broker based on their specific requirements and preferences (Fig. 3).

## 2.3   The GUI Interface of the Application

A graphical interface, integrated to the Java application, has been developed to facilitate the management of the parameters dedicated to the examples of the simulations to be executed.

The various parameters available in the graphical interface can be adjusted according to the example scenario of the cars to be simulated, the characteristics of the available field parameters are:

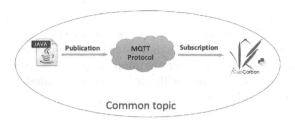

**Fig. 3.** General operation of the MQTT protocol.

✓ The possibility to choose the host server of the MQTT broker.
✓ The choice of the global topic for data sharing via MQTT.
✓ The number of cars identifiable to the MQTT topics to be generated.
✓ The possibility to choose between the default simulation file and another available Sumo file.
✓ The maximum number of iterations of a simulation.

In addition, the graphical interface offers the possibility of launching a simulation or stopping it, with two distinct buttons, dedicated to these functions.

A descriptive text, located after the buttons, indicates the current state of the simulation that is being executed (Fig. 4).

**Fig. 4.** Graphical interface of the application dedicated to accessing and sharing data of Sumo cars.

The possibility of choosing a maximum number of cars by the user gives us the interest to generate as many topics as cars to simulate. The objective of the creation of these topics is to assign, for each generated topic, a unique car which is in the course of simulation, thus, the data sharing of each car is ensured with their respective topics.

The CupCarbon tool chosen to process or visualize the coordinates of the cars shared via MQTT, needs to identify the geographical area, presented and covered by the simulation. The Libsumo library provides methods to retrieve and convert the boundary coordinates of the map proposed by Sumo, into latitude and longitude coordinates, which will allow it to be located in a geographical map of the CupCarbon IoT tool.

# 3    Digital Twin Construction With CupCarbon

The CupCarbon tool serves as an IoT simulator, integrating visualization capabilities through OpenStreetMap (OSM) maps to depict the geographical layout of a city. It employs both Python and the official SenScript.com language for scripting purposes.

By utilizing this tool, we gain access to a visualization interface for our MQTT-shared data. It facilitates the creation and manipulation of networks comprising nodes and virtual sensors. Furthermore, we can associate behaviors with these nodes through scripts, enabling seamless data accessibility via subscription to the relevant topic within our developed application.

The mechanism for data manipulation and treatment, as well as the subscription process, are effectively implemented to efficiently utilize the received data and provide real-time visualization of cars along with their updated coordinates.

## 3.1    Receiving Geographical Data and Display of Cars

The data retrieved from the topic shared with the Java application greatly facilitates data access. The Python scripts required for the retrieval and processing of the MQTT data, make the display of the geographical points of one or more cars, more dynamic as the data is extracted from the Sumo simulation, thus giving life to a digital twin of a road traffic whose source data to be collected belongs to the Sumo simulation.

The purpose of using Python in this tool is to retrieve the data emitted by the car using the MQTT protocol and then display the location of the car, in real time, on the virtual platform of CupCarbon. The data received can be saved to create complex road traffic scenarios exploring all possibilities to avoid collisions between vehicles and/or with pedestrians, or to reduce the load of road traffic in a traffic jam (Fig. 5).

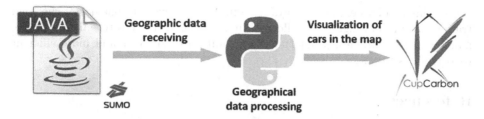

**Fig. 5.** Process of receiving data with the CupCarbon tool, for the display of cars with their respective coordinates.

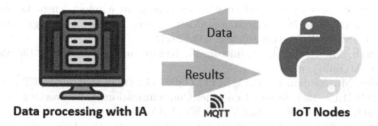

**Fig. 6.** Data processing integration scheme using artificial intelligence algorithms in future work.

## 4   Conclusion and Future Work

In this article, we have introduced a novel approach to creating a digital twin of a car by harnessing data from an existing traffic simulator. The core objective of this manoeuvre is to develop a Java application capable of extracting geographical data from simulated cars at specific time intervals. This extracted data is then seamlessly shared with the CupCarbon tool, enabling efficient processing and visualization of the available cars (Fig. 6).

In our forthcoming research, we aim to advance our methodology by integrating artificial intelligence (AI) methods into the processing and display of cars in road traffic scenarios. From a technical perspective, we plan to implement AI and machine learning algorithms using the Python interpreter under Java. The Python interpreter provides us with the flexibility to harness the functionalities and libraries available in Python, making it an ideal choice for this work. The extensive range of stable and regularly updated libraries dedicated to AI/ML in Python, validated and supported by the scientific community and developers, further strengthens this decision.

To enable AI algorithms to efficiently process and manipulate car data, we intend to utilize the MQTT protocol for sharing the data with a supercomputer equiped with a graphical processing unit. Once the processing is done, the modified data will be sent back to the Python program. This will allow for in-depth analysis and organization of the newly acquired geographic car data.

The purpose of implementing this data processing layer is to impart intelligent and autonomous behaviors to the fleet of cars in various road traffic situations. This involves exploring ways to influence the overall behavior of the road traffic system. Through these efforts, our research endeavors to enhance traffic management strategies, promoting the evolution of smarter and more efficient road traffic systems.

# References

1. Abdo, A., Wu, G., Abu-Ghazaleh, N.: Cyber-security oriented co-simulation platform for connected and autonomous driving, p. 9
2. Alsboui, T., Hammoudeh, M., Bandar, Z., Nisbet, A.: An overview and classification of approaches to information extraction in wireless sensor networks. In: Proceedings of the 5th International Conference on Sensor Technologies and Applications (SENSORCOMM'11), vol. 255 (2011)
3. Behrisch, M., Bieker, L., Erdmann, J., Krajzewicz, D.: SUMO – Simulation of Urban MObility, p. 6
4. Bender, M., Kirdan, E., Pahl, M.O., Carle, G.: Open-source MQTT evaluation. In: 2021 IEEE 18th Annual Consumer Communications & Networking Conference (CCNC), pp. 1–4. IEEE, Las Vegas, NV, USA, January 2021
5. Bounceur, A., et al.: CupCarbon: a new platform for the design, simulation and 2D/3D visualization of radio propagation and interferences in IoT networks. In: 2018 15th IEEE Annual Consumer Communications & Networking Conference (CCNC), pp. 1–4. IEEE, Las Vegas, NV, January 2018
6. Carlin, A., Hammoudeh, M., Aldabbas, O.: Intrusion detection and countermeasure of virtual cloud systems-state of the art and current challenges. Int. J. Adv. Comput. Sci. Appl. **6**(6) (2015)
7. Epiphaniou, G., Mohammad Hammoudeh, H., Yuan, C.M., Ani, U.: Digital twins in cyber effects modelling of IoT/CPS points of low resilience. Simul. Model. Pract. Theory **125**, 102744 (2023)
8. Fuller, A., Fan, Z., Day, C., Barlow, C.: Digital twin: enabling technologies, challenges and open research. IEEE Access **8**, 108952–108971 (2020)
9. Gruener, S., Koziolek, H., Ruckert, J.: Towards resilient IoT messaging: an experience report analyzing MQTT brokers. In: 2021 IEEE 18th International Conference on Software Architecture (ICSA), pp. 69–79. IEEE, Stuttgart, Germany, March 2021
10. Hadimlioglu, I.A., King, S.A.: City maker: reconstruction of cities from openstreetmap data for environmental visualization and simulations. ISPRS Int. J. Geo-Inf. **8**(7), 298 (2019)
11. Hakeem, A., Curtmola, R., Ding, X., Borcea, C.: DFPS: a distributed mobile system for free parking assignment. IEEE Trans. Mob. Comput. 1 (2021)
12. Hammoudeh, M., et al.: A service-oriented approach for sensing in the internet of things: intelligent transportation systems and privacy use cases. IEEE Sens. J. **21**(14), 15753–15761 (2020)
13. Hammoudeh, M., Newman, R.: Information extraction from sensor networks using the watershed transform algorithm. Inf. Fusion **22**, 39–49 (2015)
14. Heuer, F.M.: Scenario Generation for Testing of Automated Driving Functions based on Real Data, p. 119
15. Holm, S., Hammoudeh, M.: A comparative analysis of IoT protocols for resource constraint devices and networks. In: Proceedings of the 6th International Conference on Future Networks & Distributed Systems, pp. 616–625 (2022)

16. Horsuwan, T., Aswakul, C.: Reinforcement learning agent under partial observability for traffic light control in presence of gridlocks. In: SUMO User Conference 2019, pp. 29–9 (2019)
17. Ionescu, V.M.: The analysis of the performance of RabbitMQ and ActiveMQ. In: 2015 14th RoEduNet International Conference - Networking in Education and Research (RoEduNet NER), pp. 132–137. IEEE, Craiova, Romania, September 2015
18. Lopez-Pavon, C., Sendra, S., Valenzuela-Valdes, J.F.: Evaluation of CupCarbon network simulator for wireless sensor networks. Netw. Protoc. Algorithms **10**(2), 1 (2018)
19. Macheso, P., Manda, T. D., Chisale, S., Dzupire, N., Mlatho, J., Mukanyiligira, D.: Design of ESP8266 smart home using MQTT and Node-RED. In: 2021 International Conference on Artificial Intelligence and Smart Systems (ICAIS), pp. 502–505. IEEE, Coimbatore, India, March 2021
20. Müller, A., et al.: Towards Real-World Deployment of Reinforcement Learning for Traffic Signal Control, January 2022
21. Olaverri-Monreal, C., Errea-Moreno, J., Díaz-Álvarez, A., Biurrun-Quel, C., Serrano-Arriezu, L., Kuba, M.: Connection of the SUMO microscopic traffic simulator and the unity 3D game engine to evaluate V2X communication-based systems. Sensors **18**(12), 4399 (2018)
22. Saidallah, M., El Fergougui, A., Elalaoui, A.E.: A comparative study of urban road traffic simulators. In: MATEC Web of Conferences, vol. 81, p. 05002 (2016)
23. Sanislav, T., Mois, G.D., Folea, S.: Digital twins in the internet of things context. In: 2021 29th Telecommunications Forum (TELFOR), pp. 1–4. IEEE, Belgrade, Serbia, November 2021
24. Senyurdusev, G., Dogru, A.O., Ulugtekin, N.N.: Exploring the opportunities of open source data use in creation 3d procedural city models, vol. 1, no. 9 (2020)
25. Thangavel, D., Ma, X., Valera, A., Tan, H.X., Tan, C.K.Y.: Performance evaluation of MQTT and CoAP via a common middleware. In: 2014 IEEE Ninth International Conference on Intelligent Sensors, Sensor Networks and Information Processing (ISSNIP), pp. 1–6. IEEE, Singapore, April 2014
26. Uy, N.Q., Nam, V.H.: A comparison of AMQP and MQTT protocols for Internet of Things. In: 2019 6th NAFOSTED Conference on Information and Computer Science (NICS), pp. 292–297. IEEE, Hanoi, Vietnam, December 2019
27. Vyas, S., Tyagi, R.K., Jain, C., Sahu, S.: Performance Evaluation of Apache Kafka – a modern platform for real time data streaming. In: 2022 2nd International Conference on Innovative Practices in Technology and Management (ICIPTM), pp. 465–470. IEEE, Gautam Buddha Nagar, India, February 2022
28. Zou, S., Tao, X., Tao, B., Wu, G.: A preliminary study on the development and application of digital twin landscape architectures in the context of smart city. In: 2022 Global Conference on Robotics, Artificial Intelligence and Information Technology (GCRAIT), pp. 322–327. IEEE, Chicago, IL, USA, July 2022

# Data Summarization for Federated Learning

Julianna Devillers[1,2], Olivier Brun[2], and Balakrishna J. Prabhu[2(✉)]

[1] ISAE-SUPAERO, Toulouse, France
devillersjulianna@gmail.com
[2] LAAS-CNRS, Université de Toulouse, CNRS, Toulouse, France
{brun,Balakrishna.Prabhu}@laas.fr

**Abstract.** We explore data summarization techniques as a mean to reduce the energy footprint of Federated Learning (FL). We formulate the problem of selecting a small subset of data points that best represent the gradient of each local dataset as a submodular maximization problem and provide sufficient conditions under which the FL training is guaranteed to converge to the same global model as if the whole local datasets have been used on each client. Experimental results on IID and non-IID datasets show that this approach yields a similar accuracy as training on the full local datasets, but with a significant reduction of runtimes. There is however no clear advantage of data summarization over random sampling.

**Keywords:** Data summarization · FedAvg · convergence

## 1 Introduction

Training state-of-the-art deep learning models has an ever increasing computational cost, which doubles every few months, and consumes a lot of energy [20,21]. For instance, it is known that training the GPT-3 language model (175 billion parameters) consumed around 1,287 MWh of energy, which represents an amount of energy equivalent to the yearly consumption of 126 Danish homes and a $CO_2$ emission equivalent to 552 round-trip flights between New York and Paris [17]. It is now broadly recognized that the current race for improved model accuracy in ML is not sustainable both from an environmental and economic point of view.

There is an increasing interest in a new distributed ML paradigm called Federated Learning (FL) in which many clients collaboratively train a shared model under the orchestration of a central server, while keeping the training data private. At each round, the central server selects a subset of clients and sends them the parameters of the shared model. The selected clients train the model on

This work was partially financed by French Agence Nationale de la Recherche (ANR) through grant ANR-22-CE23-0024 (project DELIGHT).

É. Renault et al. (Eds.): MLN 2023, LNCS 14525, pp. 118–137, 2024.
https://doi.org/10.1007/978-3-031-59933-0_9

their local data and send back their gradient information to the central server for aggregation and update of the shared model. FL has a wide range of applications, especially for sensitive data for example in the fields of healthcare or finance or considering data on peoples' localization. FL has however a surprisingly large carbon footprint, which can even be up to two orders of magnitude higher than its more traditional centralized counterparts [18]. As FL is becoming more and more prevalent, the reduction of its energy footprint therefore becomes a real issue.

A first approach aims at reducing the energy spent for the communications between the clients and the central server. Gradient sparsification and gradient quantization techniques (see, e.g., [3,6,7]) have been recently proposed for that purpose. Since compression is a lossy process, the gains in terms of communication costs are usually achieved at the expense of a worse iteration complexity and there is not a good understanding of how these techniques impact the total energy consumption and whether they are worth applying in general.

Another approach, which is explored in this paper, aims at reducing the energy spent for local model updates by using data summarization techniques. The general idea is to extract a small representative set from the original local dataset on each client and to train on that smaller set. It seems reasonable to expect that training on a succinct summary rather than on the entire dataset directly can substantially reduce the energy consumption of the training phase. This raises however two main questions. The first one is: *"how to choose the data summary on each client while preserving data privacy and converging to the same global model as if the whole local datasets have been used?"*. The second one is: *"does the extra energy required for data summarization offset by the energy saved during the training phase?"*. These two questions are investigated in this paper.

## 1.1 Contributions

From a theoretical point of view, we propose to constitute the data summary on each client by choosing the data points in the local dataset which best represent the local gradient of the loss function. We show that the computation of the data summary can be cast as a submodular maximization problem or as a submodular cover problem. With the latter setup for data summarization, we provide sufficient conditions under which the FL training is guaranteed to converge to the same global model as if the whole local datasets have been used on each client.

From a practical point of view, we study the empirical test accuracy and the efficiency of the gradient matching method based on submodular maximization. Unfortunately, it is difficult to estimate the energy cost of our experiments and we use the number of rounds before convergence and the runtime instead. We note that runtime is a good proxy for the computational cost, although it does not capture the costs of communication in a real federated environment. We use the Flower toolkit to train Deep Neural Networks (DNN) on the MNIST and CIFAR10 datasets and evaluate these performance metrics in scenarios with IID and non-IID datasets. They are compared against that of the FL with the full datasets, and against random sampling. In contrast to [15], our experimental results did not show a clear advantage of data summarization over random sampling.

## 1.2   Organization

In Sect. 3, a federated learning algorithm that uses data summary is first presented, and its convergence guarantee is proven. Results from numerical experiments are detailed in Sect. 4. A discussion on our observations from these experiments as well as future work is given in Sect. 5.

Before presenting the algorithm in Sect. 3, we start by presenting in the next section the related work as well as the concepts from federated learning and submodular maximization that will be useful later on.

## 2   Preliminaries

In this section, we will summarize the two main objects of this paper. FL algorithms and data summarization techniques. First, we will present various FL algorithms and their convergence rates. This will be followed by known results in data summarization. These concepts will be used in the next section to show the convergence rate of a specific FL algorithm when it is combined with data summarization.

Federated learning (FL) [13] enables multiple actors to build a common, robust machine learning model without sharing data, thus allowing to address critical issues such as data privacy, data security, data access rights and access to heterogeneous data. It is a distributed learning paradigm with two key challenges that differentiate it from traditional distributed optimization: (1) significant variability in terms of the characteristics (hardware, data rate, etc.) of each device or client in the network (system heterogeneity), and (2) non-identically distributed data across the devices (statistical heterogeneity) which can lead to model bias during training.

In the FL setting, it is assumed that there are $K$ clients over which the data are partitioned, with $\mathcal{P}_k$ the set of indexes of data points on client $k$, with $n_k = |\mathcal{P}_k|$. The objective of the server is the following optimization problem:

$$\min_w f(w) = \sum_{k=1}^{K} \frac{n_k}{n} F_k(w) \text{ where } F_k(w) = \frac{1}{n_k} \sum_{i \in \mathcal{P}_k} f_i(w). \tag{1}$$

where $f_i(w) = \ell(x_i; y_i, w)$ is the loss of the prediction on example $(x_i; y_i)$ made with the model parameters $w$.

If the partition $\mathcal{P}_k$ was formed by distributing the training examples over the clients uniformly at random, then we would have $\mathbb{E}_{\mathcal{P}_k}[F_k(w)] = f(w)$, where the expectation is over the set of examples assigned to a fixed client $k$. This is the IID assumption typically made by distributed optimization algorithms. The IID assumption is not always satisfied in FL settings since the server has no control on how the data is distributed among the clients. The degree of heterogeneity of the data in this non-IID setting can be quantified by $\Gamma = F^* - \sum_k p_k F_k^*$, where $F^*$ (resp. $F_k^*$) is the optimal loss on the whole (resp. local) dataset and $p_k = n_k/n$ [12].

One of the first algorithm specifically designed for FL was FedAvg [13] which proposed to cooperatively train a global model. The algorithm works in discrete-time steps or rounds. At the beginning of each round, the clients receive the parameters of the global model from the central server. They then update the parameters using some standard learning algorithm on their local dataset and send the updated parameters back to the central server at the end of the round. The server averages the received model parameters and sends them to the end devices to signal the start of a new round. The pseudo-code of FedAvg is given in Algorithm 1 with $w_t$ being the parameters of current model at round $t$. The amount of computation is controlled by three key parameters: $C$ the fraction of clients that perform computation on each round; $E$, the number of training passes each client makes over its local dataset on each round; and $B$, the local mini-batch size used for the client updates.

---

**Algorithm 1.** FedAvg algorithm

---

1: **procedure** SERVER$(C, E, B)$                                  ▷ Run on central server
2:      Initialize $w_0$
3:      **for** each round $t = 1, 2, \ldots, T$ **do**
4:          $m \leftarrow \max(C\,K, 1)$
5:          $S_t \leftarrow$ random set of $m$ clients
6:          $N_t \leftarrow \sum_{k \in S_t} n_k$
7:          **for** each client $k \in S_t$ in parallel **do**
8:              $w_{t+1}^k \leftarrow$ ClientUpdate$(k, w_t)$
9:          **end for**
10:         $w_{t+1} \leftarrow \sum_{k \in S_t} \frac{n_k}{N_t} w_{t+1}^k$
11:      **end for**
12:      **return** $w_{T+1}$
13: **end procedure**

14: **procedure** CLIENTUPDATE$(k, w)$                                ▷ Run on client $k$
15:      $\mathcal{B} \leftarrow$ split $\mathcal{P}_k$ into batches of size $B$
16:      **for** each local epoch $i = 1, 2, \ldots, E$ **do**
17:          **for** batch $b \in \mathcal{B}$ **do**
18:              $w \leftarrow w - \eta \nabla \ell(w; b)$
19:          **end for**
20:      **end for**
21:      **return** $w$ to server
22: **end procedure**

---

The convergence of FedAvg has been analysed in several works [8,12,23] and the algorithm has established itself as the algorithm of choice for FL due to its simplicity and relatively low communication cost.

When there is a large statistical heterogeneity in the client datasets, the local updates in FedAvg can push the local model parameters towards the local optimum which is different from the global optimum. The global parameter models obtained from averaging can thus be different from the actual global optimum. FedProx [11] proposes to add a regularization term to the objective function at every client. The new objective is

$$h_k(w, w_t) = F_k(w) + \frac{\mu}{2}\|w - w_t\|^2. \tag{2}$$

with the second term on the RHS being the proximal term.

## 2.1 Convergence in FL

The main challenge in FL being the statistical heterogeneity, the analysis techniques used for centralized learning must be adapted for the non-IID case by adding assumptions on data dissimilarities. A number of works have studied the convergence of FL algorithms, trying to relax the less realistic assumptions. For example, the authors of [12] propose a convergence analysis for FedAvg where only a subset of clients participate in the training at each round. The authors of [24] note that the assumption taken to bound gradient dissimilarity in the work introducing FedProx [11] is quite unrealistic and propose to relax this assumption and extend the analysis to loss functions that are not smooth. However, they assume that the local client functions are $L$-Lipschitz which is also a restrictive condition on gradient dissimilarity. In Table 3 are some examples of convergence results along with the assumptions taken for FedAvg and FedProx. The explanations of the assumptions are in Table 1 and Table 2. Recall that $T$ is the total number of communication rounds, $E$ the number of local rounds between each communication round, $K$ the total number of clients, $C$ the number of clients participating in each round.

**Table 1.** Non-IID assumptions.

| Symbol | Full name | Explanation |
|---|---|---|
| $(G, B)$-BGD | Bounded gradient dissimilarity | $\mathbb{E}_k[\|\nabla F_k(w)\|^2] \leq G^2 + \|\nabla f(w)\|^2 \cdot B^2$ |
| BCGV | Bounded inter-client gradient variance | $\mathbb{E}_k[\|\nabla F_k(w) - \nabla f(w)\|^2] \leq \delta^2$ |

There are of course other significant works and approaches covering the topic of FL due to the growing interest in FL, both in industry and research. The goal of this section was however to explain some of the main concepts in FL and show the different challenges arising in the FL setup. Some of the methods covered in this section are orthogonal to each other and can be combined, some of them are only relevant in certain cases.

**Table 2.** Other assumptions and variants.

| Symbol | Explanation |
|--------|-------------|
| CVX | Each client function $F_k(.)$ is convex |
| SCVX | Each client function $F_k(.)$ is $\mu$-strongly convex |
| BNCVX | Each client function has bounded nonconvexity with $\nabla^2 F_k(x) \succeq -l \cdot \mathbf{I}$ |
| SMO | Each client function $F_k(.)$ is L-smooth |
| BLGV | The variance of stochastic gradients on local clients is bounded |
| BLGN | The expected squared norm of any stochastic gradient is bounded |
| LBG | Clients use the full batch of local samples to compute updates |
| AC | All clients participate in each round |
| Prox | Use proximal gradient steps on clients |

**Table 3.** Convergence rates.

| Method | Non-IID | Other assumptions | Variant | Rate |
|--------|---------|-------------------|---------|------|
| Yu et al. [23] | $(G,0)$-BGD | SMO; BLGV; BLGN | AC | $\mathcal{O}(\frac{1}{\sqrt{KT}})$ |
| Khaled et al. [9] | (G-B)-BGD | SMO; CVX; BLGV | AC; LBG | $\mathcal{O}(\frac{K}{T}) + \mathcal{O}(\frac{1}{\sqrt{KT}})$ |
| Li et al. [12] | $(G,0)$-BGD | SMO; SCVX; BLGV; BLGN | – | $\mathcal{O}(\frac{E}{T})$ |
| Karimireddy et al. [8] | $(G,B)$-BGD | SMO; BLGV | – | $\mathcal{O}(\frac{T(1-C/K)}{TC})$ |
| FedProx [11] | $(0,B)$-BGD | SMO; BNCVX | Prox | $\mathcal{O}(\frac{1}{\sqrt{T}})$ |

## 2.2 Data Summarization

We are given a large dataset $V$ of size $n$. The goal is to extract from $V$ a subset $S \subset V$ of data points which are most representative according to some objective function $f : 2^V \rightarrow \mathbb{R}_+$. For each $S \subseteq V$, $f(S)$ quantifies the utility of $S$. A set function $f$ is naturally associated a *discrete derivative*, also called *marginal gain*,

$$\Delta_f(e|S) = f(S \cup \{e\}) - f(S),$$

which quantifies the increase in utility obtained when adding $e \in V$ to $S$.

**Definition 1.** *A set function $f : 2^V \rightarrow \mathbb{R}_+$ is submodular if $\Delta_f(e|A) \geq \Delta_f(e|B)$ for every subsets $A \subseteq B \subseteq V$ and every data point $e \in V \backslash B$. Furthermore, $f$ is monotone if and only if $f(A) \leq f(B)$ for every subsets $A \subseteq B \subseteq V$.*

Submodular functions naturally model notions of information, diversity, and coverage in many applications. For example, let $s_{i,j}$ be the similarity between two elements $i, j \in V$ (e.g., $s_{i,j} = e^{\|i-j\|/\sigma}$). Then, the following function is submodular (non-monotone):

$$f(S) = \sum_{i \in V} \sum_{j \in S} s_{i,j} - \lambda \sum_{i \in S} \sum_{j \in S} s_{i,j},$$

where $\lambda \in [0,1]$. The first term is the traditional sum-coverage metric, while the second one penalizes similarity within $S$. Note that the function $f(S) = \sum_{i \in V} \max_{j \in S} s_{i,j}$ is another example of submodular function (see Chapter 3 of [14] for other examples).

We can distinguish two different data summarization (DS) problems of interest:

– **Submodular maximization**: the goal here is to find a summary $S^*$ of size at most $k$ that maximizes the utility, that is,

$$S^* = \operatorname{argmax}_{|S| \leq k} f(S).$$

– **Submodular cover**: the goal is to find a subset $S^*$ of data elements which achieves a target fraction of the utility provided by the full dataset, that is,

$$S^* = \operatorname{argmin} \{|S| \; : \; S \subseteq V \text{ s.t. } f(S) \geq (1 - \epsilon) f(V)\}.$$

These optimization problems are NP-hard for many classes of submodular functions [5,10,22]. However, a simple greedy algorithm proposed by Nemhauser in [16] is known to be very effective (see Algorithm 2).

---

**Algorithm 2.** Greedy algorithm

---
1: **procedure** GREEDY
2:     $S \leftarrow \emptyset$
3:     **while** $|S| < k$ **do**
4:         $v \leftarrow \operatorname{argmax}_{e \in V} \Delta_f(e|S)$
5:         $S \leftarrow S \cup \{v\}$
6:     **end while**
7:     **return** $S$
8: **end procedure**

---

**Theorem 1** ([16]). *For the submodular maximization problem of any non-negative and monotone submodular function $f$, the greedy heuristic produces a solution $S^g$ of size $k$ that achieves at least a constant factor $(1 - 1/e)$ of the optimal solution:*

$$f(S^g) \geq \left(1 - \frac{1}{e}\right) \max_{[S] \leq k} f(S).$$

*For the submodular cover problem, the approximation ratio of the greedy heuristic is $1 + \log(\max_e f(e))$, that is*

$$|S^g| \leq \left(1 + \log\left(\max_e f(e)\right)\right) |S^*|,$$

*where $S^*$ is the smallest subset (in cardinality) of $V$ such that $f(S^*) \geq (1 - \epsilon) f(V)$.*

Interestingly, [14] proposes an accelerated version of the greedy algorithm called Stochastic-Greedy that scales to voluminous datasets.

## 3   Federated Learning with Data Summary

To reduce computational and energy costs, we propose that each client extracts a data summary $S_k$ from its local dataset $\mathcal{P}_k$. Further, we also modify how updates are done by clients in each epoch. Instead of training on mini-batches and covering the whole local dataset, we propose that client $k$ perform one full-batch update per epoch on $S_k$.

The pseudo-code of the proposed FedAvg algorithm with data summary, which we call FedAVgDS, is shown in Algorithm 3. We point out two main differences with the FedAvg algorithm both of which are in the ClientUpdate procedure. First, a data summary is performed in each local epoch. And, second, the local training is performed on the whole set $S_k$ at once unlike the batch-based training in FedAvg. Since the size of $S_k$ is expected to be small, we think that dividing it into batches is not necessary and the gradient can be computed on $S_k$ in it entirety.

---

**Algorithm 3.** FedAvgDS algorithm

---

1: **procedure** SERVER$(C, E)$                 ▷ Run on central server
2:      Initialize $w_0$
3:      **for** each round $t = 1, 2, \ldots, T$ **do**
4:          $m \leftarrow \max(CK, 1)$
5:          $S_t \leftarrow$ random set of $m$ clients
6:          $N_t \leftarrow \sum_{k \in S_t} n_k$
7:          **for** each client $k \in S_t$ in parallel **do**
8:              $w_{t+1}^k \leftarrow$ ClientUpdate$(k, w_t)$
9:          **end for**
10:         $w_{t+1} \leftarrow \sum_{k \in S_t} \frac{n_k}{N_t} w_{t+1}^k$
11:      **end for**
12:      **return** $w_{T+1}$
13: **end procedure**

14: **procedure** CLIENTUPDATE$(k, w)$              ▷ Run on client $k$
15:      **for** each local epoch $i = 1, 2, \ldots, E$ **do**
16:          $S_k \leftarrow$ GreedyDataSummary$(\mathcal{P}_k, \epsilon_k)$
17:          $w \leftarrow w - \eta \nabla F_{S_k}(w)$
18:      **end for**
19:      **return** $w$ to server
20: **end procedure**

---

### 3.1   Convergence of FedAvgDS

We will prove convergence of FedAvgDS in the restricted setting of full node participation by mimicking the one of [12]. In fact, to show convergence of FedAvg, [12] modify slightly the original algorithm and assume that only one update is performed per epoch on a mini-batch. That is, the ClientUpdate procedure for the modified FedAvg in [12] is the same as that of FedAvgDS except that $S_k$ is chosen randomly and not by the GreedyDataSummary procedure.

For the convergence guarantees, we will need the same four assumptions as in [12] except that the third and the fourth ones are modified to account for the DS step as we explain below.

**A.1** $F_k$ and the $F_{S_k}$ are all $L$-smooth.

**A.2** $F_k$ and the $F_{S_k}$ are $\mu$-strongly convex.

**A.3** $\|\nabla F_k(w) - \nabla F_{S_k}(w)\| \leq \sigma_k, \forall w$. Since we do not draw samples randomly, we do not need expectations in our bounds.

**A.4** $\|\nabla F_{S_k}(w)\| \leq G, \forall w$. In FedAvg, this inequality is on the second moment of $\nabla F$ computed on the random mini-batches. Again, since we train over all of $S_k$, we need the bound on the norm of $\nabla F_{S_K}$.

**A.5** All nodes participate in all the iterations. For a random variable $X$ taking values $x_k, k = 1, \ldots K$, we have $\mathbb{E}[X] = \sum_{k=1}^{K} p_k x_k$ with $\sum_{k=1}^{K} p_k = 1$. For example, $p_k = \frac{n_k}{n}$.

**Theorem 2.** *Assume **A.1** to **A.5**. Choose $\gamma \geq max\{\frac{8L}{\mu}, E\}$ and $\eta_t = \frac{2}{\mu(t+\gamma)}$ for all $t \geq 0$. Then,*

$$F(\bar{w}_t) - F^* \leq \frac{L}{2} \frac{v}{t+\gamma}.$$

*where $v = max\{v_+, (\gamma + 1)\Delta_1\}$,*

$$v_+ = \frac{\beta^2 B + 2\beta^2 \sqrt{CD}}{\mu\beta - 1} + \frac{(2\beta\sqrt{D})^2 + \sqrt{(2\beta\sqrt{D})^4 + 4(\beta^2 B + 2\beta^2\sqrt{CD})(\mu\beta - 1)(2\beta\sqrt{D})^2}}{2(\mu\beta - 1)^2},$$

*with $C = 8(E-1)^2 G^2 + 6L\Gamma$, $D = \sum_{k=1}^{K} p_k^2 \sigma_k^2$, $B = C + D$ and $\beta\mu > 1$.*

The proof of the above theorem is in Appendix A of [4].

## 3.2  Finding a Good DS Subset

We consider two methods for extracting $S_k$ at client $k$. Both are based on distances computed from the average gradient of the loss function, and motivated from the observation that, after the local training, in an ideal setting, we would like the weights of client $k$ to be updated as:

$$w \leftarrow w - \eta \nabla_w F_k(w), \tag{3}$$

that is, by following the average gradient of the loss on the whole local dataset $\mathcal{P}_k$. Since we want the updates with DS to closely follow those without DS, one natural way is to compute $S_k$ such that the average gradient of the loss on $S_k$ is close to that of $\mathcal{P}_k$. Define

$$\nabla_w F_S(w) = \frac{1}{|S|} \sum_{i \in S} \nabla_w f_i(w), \tag{4}$$

to be the average gradient on $S$. If on $S_k$, we have that $\nabla_w F_{S_k}(w)$ is a good approximation of $\nabla_w F_k(w)$, then it will lead to a fairly similar updating of weights to that of FedAvg, and we can potentially hope to obtain convergence guarantees fairly close to those of FedAvg (i.e., without DS).

Finding such a $S_k$ corresponds to minimizing

$$\|\nabla F_k(w) - \nabla_w F_{S_k}(w)\|, \tag{5}$$

under a cardinality constraint on $S_k$. However, we are faced with the problem that (5) is not submodular, a property that is desirable if we want to call upon efficient DS algorithm we saw in Sect. 2.2.

A first method around this problem is to instead use the objective function as in Exemplar Based Clustering which can then be transformed into a monotone submodular function (see [14] for this method). Towards this end, define

$$L_k(S_k) = \frac{1}{|P_k|} \sum_{i \in P_k} \min_{j \in S_k} \|\nabla_w f_i(w) - \nabla_w f_j(w)\|, \tag{6}$$

which computes how far the points in set $P_k$ are from their closest counterparts in $S_k$. Here, the distance between two points is defined through the norm of the gradients.

The quantity in (6) can be related to a slight modification of that in (5) as follows. With $\sigma(i) = \mathrm{argmin}_{j \in S_k} \|\nabla_w f_i(w) - \nabla_w f_j(w)\|$, $A_j = \{i \in P_k, \sigma(i) = j\}$ and $\alpha_j = |A_j|$, we have:

$$L_k(S_k) = \frac{1}{|P_k|} \sum_{i \in P_k} \|\nabla_w f_i(w) - \nabla_w f_{\sigma(i)}(w)\|$$

$$\geq \frac{1}{|P_k|} \left\| \sum_{i \in P_k} (\nabla_w f_i(w) - \nabla_w f_{\sigma(i)}(w)) \right\| \tag{7}$$

$$\geq \frac{1}{|P_k|} \left\| \sum_{i \in P_k} \nabla_w f_i(w) - \sum_{j \in S_k} \alpha_j \nabla_w f_j(w) \right\|.$$

Thus, by modifying the definition of $\nabla_w F_{S_k}$ to

$$\nabla_w F_{S_k}(w) = \frac{1}{|P_k|} \sum_{i \in S_j} \alpha_j \nabla_w f_j(w), \tag{8}$$

it can be inferred that if we find a $S_k$ with $L_k(S_k) \leq \epsilon$, its norm in (5) is also at most $\epsilon$.

To transform $L_k$ into a monotone submodular function, we follow the steps in [14]. Define

$$g_k(S) = L_k(\{e_0\}) - L_k(S \cup \{e_0\}),$$

with $e_0$ a phantom point chosen such that $\max_{i' \in P_k} \|\nabla_w f_i(w) - \nabla_w f_{i'}(w)\| \leq \|\nabla_w f_i(w) - \nabla_w f_{e_0}(w)\|, \forall i \in P_k$. That is, $e_0$ is a point that is farther than any point in $P_k$ in the gradient distance. Then, $g_k$ monotone submodular.

Finally, the data summary is obtained by applying the greedy algorithm for the submodular cover problem, i.e. for a given $\epsilon_k \in \,]0, 1[$, we look for the smallest possible subset $S_k \subseteq P_k$ such that $g_k(S_k) \geq (1-\epsilon_k)g_k(P_k)$. The greedy algorithm will then find a $S_k^g$ such that $|S_k^g| \leq (1 + \log(\max_e f(e)))\,|S^*|$, where $S^*$ is the optimal subset.

It can be shown that $\|\nabla F_k(w) - \nabla F_{S_k}(w)\|$ is bounded with this method as is required in **A.4** for the convergence guarantee.

**Lemma 1**
$$\|\nabla F_k(w) - \nabla F_{S_k}(w)\| \leq \epsilon_k L_k(\{e_0\}) =: \sigma_k$$

*Proof.* From the definition of $g_k$ and the fact that $g_k(S_k) \geq (1 - \epsilon_k)g_k(P_k)$, we have

$$L_k(\{e_0\}) - L_k(S_k \cup \{e_0\}) \geq (1 - \epsilon_k) \cdot [L_k(\{e_0\}) - L_k(P_k \cup \{e_0\})].$$

Thanks to our choice of $e_0$, $L_k(S \cup \{e_0\})$ is $L_k(S)$ for any set $S$, and the above inequality reduces to

$$L_k(\{e_0\}) - L_k(S_k) \geq (1 - \epsilon_k) \cdot [L_k(\{e_0\}) - L_k(P_k)]$$

which becomes

$$L_k(S_k) \leq \epsilon_k L_k(\{e_0\}),$$

since $L_k(P_k) = 0$.

The claim follows by substituting (7) in the above inequality.     $\square$

However, this method has the drawback that it requires to find $e_0$ satisfying $\max_{i' \in P_k} \|\nabla_w f_i(w) - \nabla_w f_{i'}(w)\| \leq \|\nabla_w f_i(w) - \nabla_w f_{e_0}(w)\|$, $\forall i \in P_k$ and that minimize $\sum_{i \in P_k} |\nabla_w f_i(w) - \nabla_w f_{e_0}(w)|$ under the previous constraint. This is in itself another optimization problem. Other works usually take the 0 point as their auxiliary element [14], however the theoretic results obtained with the optimal $e_0$ can't be verified in that case.

A second method for extracting $S_k$ is to transform the problem in (5) into a monotone submodular one by putting it in the Facility Location form and to maximize:

$$\sum_{i \in P_k} \max_{j \in S_k} s(i, j) \tag{9}$$

where $s(i, j)$ is a similarity measure between points $i$ and $j$. To do so, we define $d(i, j) = \left\|\nabla_w f_i(w) - \nabla_w f_j(w)\right\|$ and $s_k(i, j) = \max_{e, e' \in P_k} d(e, e') - d(i, j)$.

This formulation is very close to the problem addressed in [15], and it becomes the same for their application to DNNs. The greedy algorithm and its accelerated versions are equally applicable to this method. In the following, we will refer to Problem (9) mainly as the CRAIG problem. In the numerical experiments, we use the last-layer approximation as in CRAIG (see [15] for details).

# 4   Numerical Experiments

All the experiments were performed using the Flower [2] Python package which provides an easy-to-use library for federated learning. Its source code is available in the GitHub repository [1]. We used a weighted variant of the cross-entropy loss and the Adam optimizer with its default parameters (except for the learning rate) for all clients in all our experiments. The dataset was either MNIST or CIFAR10.

In the legends, the terms 'ds' or 'craig' refer to the data summary obtained by solving Problem (9). The terms 'r' or 'rd' refer to random sampling. The term 'full' refers to using the full set. The number following the terms in the legend is the subset size.

For our tests, we will show the test accuracy as a function of the number of rounds or as a function of the runtime. Indeed, the communication cost depends on the number of rounds, while the computational cost depends on the running time.

## 4.1   IID Datasets

Our first tests were done using a Quad Core Intel Core i7-7700 CPU. We first used MNIST dataset, splitting the training dataset into 10 datasets and splitting each of that dataset into a training and a validation set with a ratio of 90/10%. Therefore, each of our ten clients had a training set of 4500 images and a validation set of 500 images. We gave the MNIST test set to the server and used the implemented FedAvg [13] strategy of Flower.

On MNIST, we compared extracting a data summary using the CRAIG method, with random sampling, or the full dataset, with the same initial model parameters. The results presented here are for the accuracy on the test set. For Figs. 2 and 3, the subset size is of 500 images while it is of 200 for Fig. 1. The FL training included $R = 10$ rounds of training where at each round the server sends the current global parameters to the ten clients, the clients locally train the model for $E = 5$ epochs on the required dataset (full, random or data summary) and then sends its updated parameters to the server that aggregates all the responses to update the global model and so on.

Figure 1 shows that using subsets greatly accelerates training in terms of execution time when compared to the full dataset in our setup. However, this requires more rounds. In that configuration, we were able to reach a test accuracy of 0.980 in 8 rounds taking 18 min with the data summaries while it took 1 h22 to complete 2 rounds of training on the full dataset to achieve a test accuracy of 0.976. In the FL setup, using submodular maximization seems to outperform random sampling. In 19 min corresponding to 10 rounds, the model was only able to reach an accuracy of 0.933 with random sampling. This comes from the fact that data selection time is much smaller compared to the whole training process.

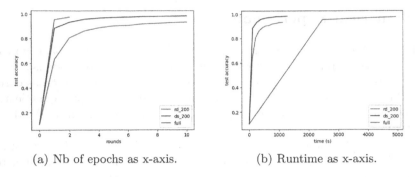

(a) Nb of epochs as x-axis.    (b) Runtime as x-axis.

**Fig. 1.** Test accuracy compared between random sampling, CRAIG sampling or the full dataset for MNIST IID data. $E = 5$, $R = 10$, $\eta = 1e - 3$. Full set size is 50000 and subset size is 0.0044.

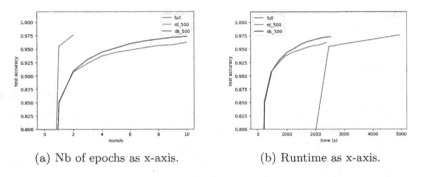

(a) Nb of epochs as x-axis.    (b) Runtime as x-axis.

**Fig. 2.** Test accuracy compared between random sampling, CRAIG sampling or the full dataset for MNIST IID data. $E = 5$, $R = 10$, $\eta = 1e - 3$. Full set size is 50000 and subset fraction size is 0.011. FL setup.

Given that the communication rounds are very costly in FL, we tried to compensate the increased number of rounds by increasing the learning rate $\eta$. Figure 3 shows that increasing the learning rate did not really improve the training with the full dataset. For the data summaries however, increasing the learning rate in that configuration enabled the model to reach an accuracy of 0.96 in 2 rounds and 9 min when it took it 6 rounds and 35 min with the previous learning rate. In only 1 epoch, the model trained on the whole dataset was able to reach an accuracy of 0.954 in 41 min. Training on subsets therefore seems to greatly decrease the local training time but we were still not able on IID data to achieve a given accuracy in the same number of rounds as the full dataset training. This could however be possible on certain tasks or by adjusting some hyperparameters such as the learning rate $\eta$, the rate of subset selection (every $E$ epochs in that configuration) or the subset size.

Data subset selection seems far more relevant here, especially as the training takes more time. This shows that data subset selection, using random sampling or CRAIG sampling, can significantly speed up training for bigger models (with

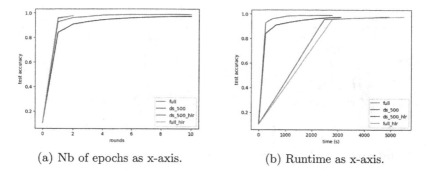

(a) Nb of epochs as x-axis.          (b) Runtime as x-axis.

**Fig. 3.** Test accuracy compared between random sampling, CRAIG sampling or the full dataset for MNIST IID data. $E = 5$, $R = 10$, $\eta = 1e - 2$ for the high learning rate (_hlr) and $\eta = 1e - 3$ for the other. Full set size is 50000 and subset fraction size is 0.011.

long training time) or for devices with limited system resources. This could help reduce the energy consumption in the case where reducing the dataset does not induce a drop in training accuracy and therefore the need for more computation rounds. Several hyperparameters are to be tuned for that, including the subset size, the number of local epochs and even the learning rate.

## 4.2   Non-IID Datasets

In this part we tested two configurations. In the first one, 10 Flower clients run on a single machine with an Intel 5218R 2.1G CPU (20 cores) and one GPU RTXA6000. The datasets are divided as in the previous section. In the second configuration, 4 Flower clients run on the same machine, 2 on one RTXA6000 GPU and 2 on another RTXA6000 GPU. The dataset is split so that each client has images corresponding to 5 labels: 1 to 5 for the first, 6 to 10 for the second, only the even-numbered for the third and the uneven for the last one. This second configuration is therefore less non-IID than the first one. Each local dataset contains 12500 images while they contain 5000 images in the first configuration. On both configurations, the server is run on another machine (Intel E5-2695 v3 2.3G CPU). As the computing time is reduced with the use of GPU, we also decided to use ResNet-18 as this model is the one commonly used in the other works for tests on CIFAR-10. [15,18,19].

As shown in Fig. 4, the configuration with 10 clients, using CRAIG for data-summary is far too time-consuming compared with random sampling or even using the full set, as mentioned earlier in this report. However, random sampling seems to perform better than using the full set, even if we reduce the number of training epochs for each round to $E = 1$. This probably comes from the chosen data distribution. As we are in a non-iid configuration, each local model will be updated towards a local optimum which is not the same as the global one. If we let each client make a large number of local updates, their local model will be

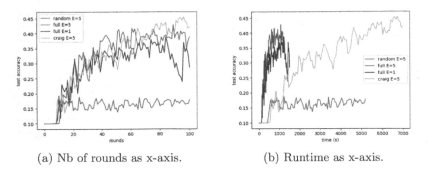

(a) Nb of rounds as x-axis.          (b) Runtime as x-axis.

**Fig. 4.** Test accuracy compared between random sampling, CRAIG sampling or the full dataset for CIFAR-10 non-IID data, using ResNet-18. $R = 100$, $\eta = 1e - 3$, $\mu = 0.01$. Subset fraction size is 0.11. FL setup with 10 clients on 1 GPU.

updated towards that local optimum which can be self-defeating for our training. This seems to be what happens in the case where $E = 5$ where the model fails to learn due to an excessive number of local updates.

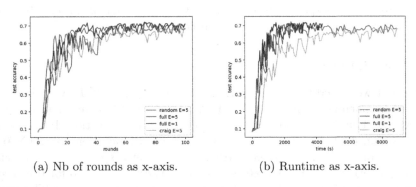

(a) Nb of rounds as x-axis.          (b) Runtime as x-axis.

**Fig. 5.** Test accuracy compared between random sampling, CRAIG sampling or the full dataset for CIFAR-10 non-IID data, using ResNet-18. $R = 100$, $\eta = 1e - 3$, $\mu = 0.01$. Subset fraction size is 0.11. FL setup with 4 clients on 2 GPU.

In the configuration with 4 clients as shown in Fig. 5, the conclusion for the CRAIG data summarization is even clearer. As each local dataset here contains 12500 images, with 1250 in the validation set and the rest in the training set, extracting a subset containing 1250 images (a 0.11 fraction as before) is even more time-consuming. Full set strategies perform better than random and CRAIG sampling in terms of test accuracy at each round, although random and CRAIG sampling soon catch up.

The results in both Fig. 4 and 5 suggest that there are cases where using the full set even for only one epoch is not necessary, making the use of subsets relevant.

**Frequency of Subset Selection.** In the previous section, using CRAIG was too time-consuming. One way of limiting this is to play on the frequency with which the subset is selected. On all our previous experiments, we selected a new subset at each round but we can also decide to choose a new subset each $p$ rounds. In this part, we use the same data partition as before with 10 clients, each having a training set of size 4500 representing 2 different labels for CIFAR-10. We distribute those 10 clients on 2 GPUs.

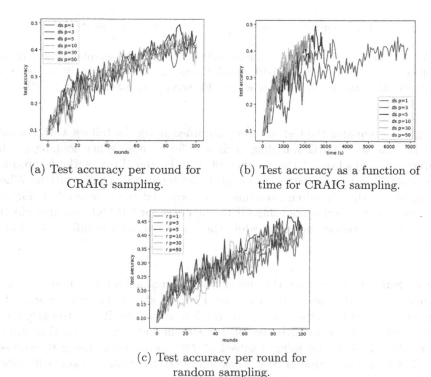

(a) Test accuracy per round for
CRAIG sampling.

(b) Test accuracy as a function of
time for CRAIG sampling.

(c) Test accuracy per round for
random sampling.

**Fig. 6.** Test accuracy for random sampling and CRAIG sampling for different values of the period $p$ for extracting the subsets. CIFAR-10 non-IID data, using ResNet-18. $R = 100$, $\eta = 1e - 3$, $\mu = 0.01$. Subset fraction size is 0.11. FL setup with 10 clients on 2 GPU.

Figure 6 shows the test accuracy obtained for different values of the period $p$ for random and CRAIG sampling. Interestingly enough, selecting the subset less frequently does not lead to a significant drop in precision. Meanwhile, this indeed allows us to save some computing time on CRAIG sampling.

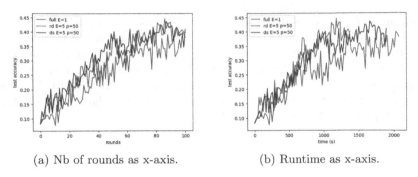

(a) Nb of rounds as x-axis.                    (b) Runtime as x-axis.

**Fig. 7.** Test accuracy compared between full set or random sampling and CRAIG sampling with $p = 50$ for CIFAR-10 non-IID data, using ResNet-18. $R = 100$, $\eta = 1e - 3$, $\mu = 0.01$. Subset fraction size is 0.11. FL setup with 10 clients on 2 GPU.

Figure 7 compares the test accuracy obtained using the full set with the one using random or CRAIG sampling each $p = 50$ rounds. When looking at the accuracy as a function of the number of rounds, the three methods produce close results. Craig sampling seems slightly better in this configuration. When looking at the time, random sampling is as expected the quickest but, for the first time in our experiments using GPU, the curve for CRAIG sampling closely follows that of random sampling while the one for using the full set is clearly below.

**Size of Subset.** Of course, the size of the selected subset is a hyperparameter that can greatly impact the training. Figure 8 shows the test accuracy for different values of the subset size $S$ for random sampling. If $S$ is too large, the local model will move towards the local optimum which can be detrimental to the training. This is why using a subset of 2250 outperforms using the full set ($S = 4500$). A large value of $S$ will also increase the runtime, especially when using submodular maximization for subset selection. If $S$ is too small, the training will take much longer in terms of rounds which will result in an increased communication cost. This is what would happen for $S = 500$ where the model learns and eventually achieves the accuracy reached by bigger subsets but this takes more rounds. A balance has to be found. In addition to that, the optimal value for $S$ depends on the chosen value of $E$ or even of the learning rate $\eta$.

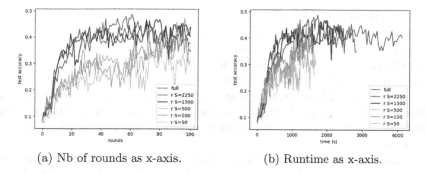

(a) Nb of rounds as x-axis.          (b) Runtime as x-axis.

**Fig. 8.** Test accuracy compared for various sizes of the subset selected by random sampling for CIFAR-10 non-IID data, using ResNet-18. $R = 100$, $\eta = 1e - 3$, $\mu = 0.01$. Subset fraction size varies. FL setup with 10 clients on 2 GPU.

## 5  Discussion and Future Work

Based on the results of our experiments, we can clearly identify two cases in which the use of subsets of local datasets seems relevant. In both however it is not clear how much improvement can be made by choosing the subset using a DS method rather than randomly selecting one.

The first case concerns clients with limited computational resources. In those cases, the training can take a very long time (the straggler problem) due to clients that are holding back other faster one. In some cases, it is important to also being able to include stragglers in the training loop, especially if the data points of the stragglers don't appear sufficiently in other datasets. Using subsets instead of the full set can significantly decrease the training time, particularly if those clients have large datasets, and therefore help them participate in the FL training.

The second case concerns clients with large datasets that are not representative of the whole data distribution. Basically, this means that their local optimum for the model's parameters is different from the global optimum. To prevent them from converging too much towards their local optimum, hyperparameters have to be chosen wisely. This can be done for example by reducing the learning rate or increasing the batch size in order to reduce the number of local updates. However, this is not always the best choice for example if the training works better with the initial hyperparameters for other clients with smaller datasets and more generally if we can achieve the same performance simply by going through a smaller number of points of their datasets. This is our interpretation of what happens in Sect. 4.2 where using subsets leads to better results than using the full set for 1 epoch.

Selecting a subset also requires tuning some parameters including the size of the subset and the selection frequency. In our case, reducing the selection frequency did not lead to a drop in accuracy while, for the data summarization method, it helped reduce the running time as shown in Sect. 4.2. Considering the size of the subset, a balance has to be found as explained in part 4.2. Tuning those parameters can be done locally in order to avoid additional communication rounds. It could of course result in an additional computational cost but this could easily be offset for long and costly trainings, for example in a FL hyperparameter tuning context. However, it is important to point out that the use of subsets also influences the choice of certain hyperparameters, such as the number of epochs for local training in each round. More indirectly, the choice of other parameters, such as the learning rate, may depend on whether or not subsets are used.

Our approach focused on finding a subset approximating the full set gradient for the loss. Our study did not show clear improvements in the test accuracy or the loss. Considering the longer runtimes, this means that using submodular maximization rather than random sampling was not relevant in our case. This could be linked to the datasets we used or the way we distributed them between clients. Indeed, in our experiments, each client's local dataset would actually be quite homogeneous, with 5000 images of only 2 labels resulting in a lot of redundancy. Based on that, it is not surprising that random performed well. One might think that results would be different on more complex datasets, but this has not been studied here and is part of our future work,

Another avenue of investigation is to determine the actual energy consumption of the algorithms with and without energy and determine the tradeoff between the accuracy and the energy consumption.

# References

1. Flower github repository (2023). https://github.com/adap/flower. Accessed 30 Aug 2023
2. Beutel, D.J., et al.: Flower: a friendly federated learning research framework. arXiv preprint arXiv:2007.14390 (2020)
3. Cui, L., Su, X., Zhou, Y., Zhang, L.: ClusterGrad: adaptive gradient compression by clustering in federated learning. In: 2020 IEEE Global Communications Conference 2020, pp. 1–7 (2020)
4. Devillers, J.: Data summarization methods for energy efficient federated learning. Master's thesis internship report, ISAE-SUPAERO (2023). https://github.com/juliannadvl/FL-DS/blob/main/rapport_stage.pdf
5. Feige, U.: A threshold of ln n for approximating set cover. J. ACM **45**(4), 634–652 (1998). https://doi.org/10.1145/285055.285059
6. Haddadpour, F., Kamani, M.M., Mokhtari, A., Mahdavi, M.: Federated learning with compression: unified analysis and sharp guarantees. In: PMLR, vol. 130, pp. 2350–2358 (2021)
7. Jiang, P., Agrawal, G.: A linear speedup analysis of distributed deep learning with sparse and quantized communication. In: NeurIPS (2018)
8. Karimireddy, S.P., Kale, S., Mohri, M., Reddi, S., Stich, S., Suresh, A.T.: Scaffold: stochastic controlled averaging for federated learning. In: III, H.D., Singh, A. (eds.) Proceedings of the 37th International Conference on Machine Learning. Proceedings of Machine Learning Research, vol. 119, pp. 5132–5143. PMLR (2020). https://proceedings.mlr.press/v119/karimireddy20a.html

9. Khaled, A., Mishchenko, K., Richtárik, P.: Better communication complexity for local SGD. CoRR abs/1909.04746 (2019). http://arxiv.org/abs/1909.04746

10. Krause, A., Guestrin, C.: Near-optimal nonmyopic value of information in graphical models. In: Proceedings of the Twenty-First Conference on Uncertainty in Artificial Intelligence, UAI 2005, pp. 324–331. AUAI Press, Arlington (2005)

11. Li, T., Sahu, A.K., Zaheer, M., Sanjabi, M., Talwalkar, A., Smith, V.: Federated optimization in heterogeneous networks (2018). https://doi.org/10.48550/ARXIV. 1812.06127. https://arxiv.org/abs/1812.06127

12. Li, X., Huang, K., Yang, W., Wang, S., Zhang, Z.: On the convergence of FedAvg on Non-IID data (2020)

13. McMahan, B., Moore, E., Ramage, D., Hampson, S., Aguera y Arcas, B.: Communication-efficient learning of deep networks from decentralized data. In: Singh, A., Zhu, J. (eds.) Proceedings of the 20th International Conference on Artificial Intelligence and Statistics. Proceedings of Machine Learning Research, vol. 54, pp. 1273–1282. PMLR (2017). https://proceedings.mlr.press/v54/mcmahan17a. html

14. Mirzasoleiman, B.: Big data summarization using submodular functions. Ph.D. thesis, ETH Zurich (2017)

15. Mirzasoleiman, B., Bilmes, J., Leskovec, J.: Coresets for data-efficient training of machine learning models. In: Proceedings of the 37th International Conference on Machine Learning, ICML 2020. JMLR.org (2020)

16. Nemhauser, G.L., Wolsey, L.A., Fisher, M.L.: An analysis of approximations for maximizing submodular set functions - I. Math. Program. 14, 265–294 (1978)

17. Patterson, D., et al.: Carbon emissions and large neural network training (2021)

18. Qiu, X., et al.: A first look into the carbon footprint of federated learning (2023)

19. Reddi, S.J., et al.: Adaptive federated optimization. In: International Conference on Learning Representations (2021). https://openreview.net/forum? id=LkFG3lB13U5

20. Schwartz, R., Dodge, J., Smith, N.A., Etzioni, O.: Green AI (2019)

21. Strubell, E., Ganesh, A., McCallum, A.: Energy and policy considerations for deep learning in NLP. In: Proceedings of the 57th Annual Meeting of the Association for Computational Linguistics, pp. 3645–3650. Association for Computational Linguistics, Florence (2019). https://doi.org/10.18653/v1/P19-1355. https://aclanthology.org/P19-1355

22. Weinberg, M.: Lecture notes in advanced algorithm design - lecture 7: submodular functions, lovász extension and minimization (2022). https://www.cs.princeton. edu/~hy2/teaching/fall22-cos521/notes/SFM.pdf. Accessed 19 Apr 2023

23. Yu, H., Yang, S., Zhu, S.: Parallel restarted SGD with faster convergence and less communication: demystifying why model averaging works for deep learning (2018)

24. Yuan, X.T., Li, P.: On convergence of FedProx: local dissimilarity invariant bounds, non-smoothness and beyond (2022)

# ML Comparison: Countermeasure Prediction Using Radio Internal Metrics for BLE Radio

Morgane Joly[1], Éric Renault[2(✉)], and Fabian Rivière[1]

[1] NXP Semiconductors, 2 esplanade Antone Philips, 14000 Caen, France
[2] LIGM, Univ. Gustave Eiffel, CNRS, ESIEE Paris, 93162 Noisy-le-Grand, France
eric.renault@esiee.fr

**Abstract.** The reliability of low-power wireless communications is being challenged by the proliferation of Internet of Things (IoT) devices. In an increasingly dynamic context, new countermeasure management is needed to make the IoT network more flexible. This paper proposes a comparison between three machine learning (ML) algorithms to predict the next countermeasure to be applied during the next wake-up slot of a BLE receptor. We evaluate the ratio between performance and stability of the solution. The best technique that emerged is the bagged tree, which predicts the future countermeasure to be implemented with an accuracy of 99.7%.

**Keywords:** Cognitive Radio · Machine Learning · RF Countermeasures · BLE

## 1 Introduction

With the Internet of Things (IoT) boom, the spectrum is becoming increasingly crowded, increasing costs and are forcing innovation to focus on the energy efficiency of the devices without compromising on the performance already achieved. To address this issue, many wireless protocols have been created in recent last years to address this issue. The Bluetooth Low Energy (BLE) protocol, which operates in the 2.4 GHz band for its lightweight communication which allows to extend the battery life of devices. It is impacted by the presence of Wi-Fi, which is also active in this band (see Fig. 1). The increase in communication provokes spacial overloads, which impacts the quality of service (QoS) of any wireless protocol [2].

Even if coexistence techniques, such as adaptive frequency hopping (AFH), are used to reduce the impact of interference, over-consumption is inevitable due to packet collision resulting in multiple re-emission. Latency increases during device discovery procedure and is correlated with the number of interfering devices of the same protocol, and large sensor networks are degraded by Wi-Fi.

The paper contribution lies in the evaluation and the comparison of the performances of three machine-learning (ML) based algorithms. The aim of these

© The Author(s), under exclusive license to Springer Nature Switzerland AG 2024
E. Renault et al. (Eds.): MLN 2023, LNCS 14525, pp. 138–147, 2024.
https://doi.org/10.1007/978-3-031-59933-0_10

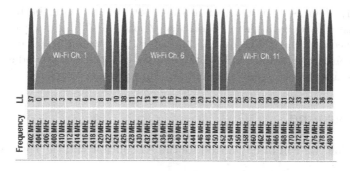

**Fig. 1.** Superposition of Wi-Fi and BLE communication channels in the 2.4 GHz band

algorithm is to predict the best countermeasure among three – change channel, restrict Automatic gain control (AGC) index, no countermeasure – for the next packet reception according to the last packet. To address the AGC index restriction possibility we show that restricting the AGC for a subset of packets contributes to improve their reception.

Regarding the countermeasure manager the results show that with few metrics (about 200) we can achieve a prediction of the best countermeasures among two with an accuracy of 99.7% using a bagged tree. We also show the interest of restraining the AGC index for identified conditions to save 94% of the packets under the identified conditions.

This paper is organised as follows. Related works are presented in Sect. 2. Section 3 introduces knowledge to ease the reading of Sect. 4, which presents the environmental setup used to collect and to pre-process the data. Results are presented and discussed in Sect. 5. Finally, we conclude in Sect. 6.

## 2  Related Work

Integration of artificial intelligence in the domain of the non-collaborative coexistence of wireless protocols is a well-documented subject.

The intelligent channels assignment search to improve the adaptive frequency hopping usually with a whitelisting or blacklisting of channels. To the best of our knowledge, Nikoukar and et al. [4] are the only team to propose a model (LSTM-based) that addresses the specific issue of Wi-Fi interference prediction on BLE channels.

The detection of interference can contribute to a better detectability and allow the reconstruction of packets as well as malicious activity surrounding the device. As T. Kikuzuki and et al. [3], which propose an algorithm to identify nearby protocol to improve detection sensitivities while there is emission overlap tanks to a new layout of signal decoding. The study [1] differentiate the events that caused the loss of the packet and adapt the emission of packets to the spectral environment.

Natively, devices cannot detect nearby protocol to apply the best counter-measure associated with them. Wang and Zhang in [6] use a decision tree to take intp account the MAC layer in the spectrum sensing scheme to design a new emission strategy and improve the quality of service of secondary users by wisely choosing the emission time and avoiding primary users.

## 3    Background

In this section, we will briefly introduce the AI concept and the specifics of BLE radio.

### 3.1    Artificial Intelligence Background

We present here the three algorithms compared and give an insight into how they work. Moreover, the algorithmic complexity of the solution presented is also provided. Let $n$ be the number of samples and $p$ be the number of features.

*Decision Tree.* The decision Tree algorithm is a classic ML algorithm used for classification problems, known for its great generalisation capabilities and for working well with large datasets. The general training complexity of this algo-rithm is $O(n \times p \times \log(n))$.

*Bagged Tree.* This kind of approach combines several decision trees (called weak learners) using bootstrap aggregation method to form a stronger estimator. The trees are trained on a different fraction of the training set, the final output is the most popular output of all the trained models. The general training complexity of this algorithm is $O(n^2 \times p \times \log(n))$.

*Support Vector Machine (SVM).* SVM is a family of algorithms originally designed to discriminate between two classes, it is based on the search for opti-mal hyperplanes separating the data while maximising the margin to facilitate the generalisation capacity. The general training complexity of this algorithm is $O(n^2 \times p)$.

### 3.2    Radio Frequency Background

Before explaining the context of the study, some terms need to be defined: Met-rics can be collected during packet reception, they characterise the evolution of the channel environment, e.g. observe the arrival of an interferer.

*Fast Link Estimators: Received Signal Strength Indicator (RSSI), The Signal to Noise Ratio (SNR) and Link Quality Indicator (LQI).* Three complementary hardware-based metrics of the link quality between two radios. RSSI provides an estimate on the average power in a channel throughout the reception of a packet, all signals included, while LQI and SNR quantify the signal-to-noise ratio, i.e.

the capacity of the receiver to decode the signal despite ambient noise. SNR gives the margin of the signal over the noise and the ability of the demodulator to detect the bits.

They can be used to quickly estimate the radio receiver state. However, they are subject to various biases that make them poor resources for determining the presence of interferers.

*Cyclic Redundancy Check (CRC).* The CRC is a word composed of a few bits calculated by the emitting radio from the payload part of the send packet. It is appended to the end of the packet before its emission. When the receiver decodes the packet, this word is recalculated with the received payload and compared with the received CRC. The receiver can then determine if the received packet has been corrupted.

*Access Address Found (AAF).* This binary metric provides a confirmation of the detection of packet start. There are two scenarios that can lead to an access address not being detected:

– Wanted signal strength is too weak for the receiver despite missing interferer.
– There is a too powerful interferer at the beginning of the packet reception.

*I/C.* This simulation input data represents the ratio of interference power to wanted signal power at the antenna level during the simulation. When the wanted signal power is sufficiently higher than the interferer signal power, the interferer is not differentiable from the wanted signal so, in this case, the interferer cannot disturb the packet reception.

*Packet Reception Status.* This indicator is the logic combination of AA found and CRC. It allows to summarise the packet reception condition (see Table 1).

**Table 1.** Creation of packet reception status indicator

|          | Good reception (GR) | Bad reception (BR) | No detection (ND) |
|----------|---------------------|--------------------|-------------------|
| CRC      | 1                   | 0                  | 0                 |
| AA Found | 1                   | 1                  | 0                 |

*Automatic Gain Control (AGC) Index.* The AGC is used to avoid saturation during the payload reception by adjusting the receiver gain according to the received signal strength. The index is set at the beginning of the packet reception, during the preamble, and frozen during the payload to avoid corruption of the bit decoding. The AGC is therefore not set to sustain an increase in signal power during the payload reception, which occurs when an interferer level creates receiver saturation and signal filtering is not efficient to eliminate the interferer.

The higher the interferer, the lower the receiver gain and consequently the lower the AGC index value. Later in this paper the terms of under-restricted for an AGC index value higher than needed and over-restricted for an AGC index value lower than required, will be used.

If the interference arrives after the AGC index freeze then the signal will be saturated and the packet will be corrupted (see Fig. 2). The packet re-emission process will be triggered, causing an over consumption of energy compared to a well received packet on the first attempt.

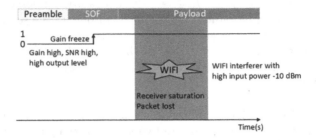

**Fig. 2.** Packet loss scenario due to Wi-Fi arriving after AGC index freezes

We observe two options of improvement:

- In case interferer arrives after AGC index freeze, for some packets, a more restrictive AGC index guaranty a good reception.
- In case the AGC index extra adjustment cannot reduce the impact of the interferer, channel hopping is required and operation on busy channel must be avoided.

In order to improve the current countermeasure systems, we hypothesised that future reception conditions can be predicted from past reception and hence future countermeasures. The Sect. 4 explains how we set up the experimentation and the Sect. 1 shows the result analysis.

# 4    Countermeasure Prediction Methodology

## 4.1    Delimitations and Scope

The proposed tool is experimented in the context of a BLE protocol interfered by a Wi-Fi interferer, it does not take into account other possible interferers (Bluetooth, BLE, microwave oven) due to the need to finely control the environment. However, this tool can be extended to other low-power protocols and other interference configurations will be the object of future experiments. The experimental environment setup consists of a BLE receiver simulated using Matlab (v2020b) and Simulink modelling. We assume that there is only one Wi-Fi interferer at 12 MHz offset from the wanted signal. To reduce the complexity of the problem, the frequency hopping is disable, and it is assumed that emission and reception remain on the same channel throughout the experiment.

## 4.2   Dataset Creation Method

In this section, the method used to built the datasets is presented. To reduce the simulation time and to obtain sufficient data to train the models, we choose to create synthetic signals from collected packets generated by semi-random series.

**Packet Simulation.** A packet is generated by setting the power of the desired signal (in the range from $-100$ dBm to $0$ dBm with a step of 2 dB) and the interferer power (in the range from $-120$ dBm to $0$ dBm with a step of 2 dB). The arrival of the interferering signals is controlled. Two use cases are considered:

– The interferer is at the start of a BLE packet reception (before AGC freeze)
– The interferer arrives during the payload (after AGC freeze)

Due to anticipation of reduced memory, computational resources and energy, and to accelerate the prediction speed, the number of inputs is reduced to the bare minimum. We have chosen to pre-process the signal by collecting only three representative data at different intervals within the packet under reception, at the beginning of the preamble, at the payload mid and before the CRC (see Fig. 3.

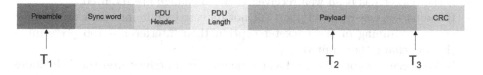

**Fig. 3.** Intervals of measures during packet reception

**Synthetic Signal Creation.** The input signal is generated from the collected packet created by a semi-random series of packets where the interferer is present or absent, corresponding to simplified Wi-Fi duty cycles. We create reality-related browsing behaviour signal based on the duty cycle described by Rajesh Palit's analysis of Wi-Fi in [5].

The packets from the data generation are sorted according to the I/C ratio, which is an insight into the effect of the interferer on the signal. The greater is the I/C ratio, the more the wanted signal will be impacted (see Table 2).

Synthetic signals are created following the chosen Wi-Fi cycle ratio by alternating a number of undisturbed packets (corresponding to the No interferer class) and packets interfered by low, medium or high power interferers (see Fig. 4).

The proportion of packets with interferer arriving after the AGC index freeze is controlled by a percentage set from 0% to 100% with a step of 20%.

**Table 2.** Threshold of I/C which determines the impact of interferer on BLE Signal

| Level | No interferer | Low | Medium | High |
|-------|---------------|-----|--------|------|
| I/C | $x < -20$ | $-20 < x < 0$ | $0 < x < 20$ | $20 < x < 35$ |

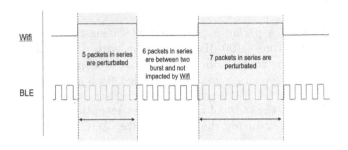

**Fig. 4.** Example of packet arrangement to recreate Wi-Fi pattern

**Data Labelisation.** To label the packet we use the reception status indicator. Only two countermeasures are available, AGC index restriction and channel switching (see Table 3):

- If the packet has been well received then no change are required.
- If the packet was received poorly or not detected while interferer was present at the beginning of the packet reception then interferer is too powerful, a channel change is required.
- If the packet was bad received while the interferer arrives after the AGC index freeze then if the AGC index has been more restricted, the packet could have been well received, a restricted AGC is required.
- Else there is no improvement that can be done so no specific countermeasure are applied.

**Table 3.** Constitution of the labels

|    | Before AGC freeze | After AGC freeze | No interferer |
|----|-------------------|------------------|---------------|
| GR | no change | no change | no change |
| BR | channel change | index AGC restriction | no change |
| ND | channel change | no change | no change |

**Train and Test Dataset.** Training and test sets are realised from six different wanted signal powers: $-47$ dBm, $-37$ dBm, $-27$ dBm, $-17$ dBm, with interferer power set to four levels. We set the sliding window in the range 10. A 13-fold cross-validation to partition the input is used for all training.

## 4.3   Validation Dataset

To validate the algorithms performances and there capacity of generalisation, a validation dataset is created with wanted signal powers from −69 dBm to −19 dBm with a step of 20 dB, so that the packets used in the validation set are different from those used in the training and test sets.

# 5   Performance Evaluation and Analysis

This section provides the results of training and validation of the ML model, the bagged tree, the decision tree and the SVM.

## 5.1   Effect of the Restrained AGC Index on GR_BR Class

To observe the interest of restraining the AGC index we have selected a subset of well received packets when the interferer is present since the start of emission but poorly received when the interferer arrives after the AGC freeze with different frequency offsets. These 80 packets were re-emitted in a controlled context to observe the potential improvement of this countermeasure.

By forcing the AGC index of the *After* version packets of this class, we obtain an improvement of 61% on the selected packet sample (see Fig. 5).

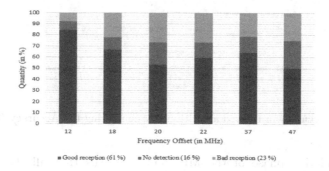

**Fig. 5.** Recovered packets reception status after AGC index forcing (packets originally missed due to interferer arrival after AGC freeze, then AGC forced to the index value got when interferer occurrence before wanted packet)

Without the AGC forcing at 12 MHz frequency offset 15% of packets are not received and with AGC forcing only 2% of packets are not received. On average, at each frequency offset, 62% of the packets are well received.

The result of the remission of different packets with a restriction of the AGC index confirms the interest of integrating the AGC restriction in the counter-measure panel. It could allow a better detection of packets in powerful interferer situation.

## 5.2   Next Slot Countermeasure Prediction

During the validation phase of models from −69 dBm to −19 dBm with a step of 20 dB signal, bagged tree show a mean accuracy of 99.6% ± 0.07%, Decision Tree show a mean accuracy of 98.7% ± 0.18%, SVM show a mean accuracy of 99.7% ± 0.01% (see Table 4).

**Table 4.** Experiment result

| Algorithm tested | Bagged Tree | Decision Tree | SVM |
|---|---|---|---|
| Result of classification | 99.7% | 98.5% | 83.3% |
| Confidence Interval | 0.17% | 0.69% | 3% |

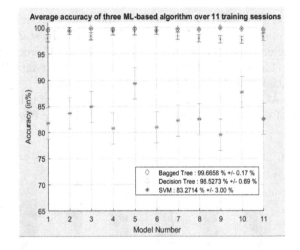

**Fig. 6.** Accuracy of trained algorithms by comparing the Accuracy (in %)

The validation phase of the ML models reveals that all the algorithms tested perform well in predicting the next packet countermeasures, especially the Bagged Tree, which is the most efficient and stable, as shown in Table 4 and Fig. 6. Since the validation dataset is decorrelated from the trainning and test sets, we can assume that the algorithm can generalise well to new data. Nonetheless, the high performances are certainly realated to the ideal conditions of data collection and signal generation. Future experiments will be dedicated to testing our assumptions under real conditions.

## 6   Conclusion

Interference has a strong impact on wireless communications. Little research has been conducted on the use of specific receiver-internal metrics generated during

BLE packet reception to predict countermeasures for future receiver wake-up slots. Here, we have presented three machine learning-based algorithms that can be used to predict countermeasures, namely providing a more restrictive AGC or changing the channel, to prepare for the reception of the n + 1 packet. The experiment has shown that bagged tree is the best one in terms of accuracy and stability with 99.7% $\pm$ 0.17 % of accuracy. We have shown that it is possible to improve the reception of certain packets by selecting a more restrained AGC index.

## 7    Future Work

This paper presents results under ideal conditions of generation. The aims for future work are to test our assumption in real time on simulation and then to test our countermeasure management on hardware with real interference. We also plan to add an emission delay to enrich our range of countermeasures.

## References

1. Eu, Z.A., Lee, P., Tan, H.P.: Classification of packet transmission outcomes in wireless sensor networks. In: 2011 IEEE International Conference on Communications (ICC), pp. 1–5 (2011). https://doi.org/10.1109/icc.2011.5962637
2. Feng, W., Arumugam, N., Hari Krishna, G.: Impact of interference on a Bluetooth network in the 2.4 GHz ISM band. In: The 8th International Conference on Communication Systems 2002, ICCS 2002, vol. 2, pp. 820–823 (2002). https://doi.org/10.1109/ICCS.2002.1183244
3. Kikuzuki, T., Wada, A., Hamaminato, M., Ninomiya, T.: Automatic standard classification method for the 2.4 GHz ISM band. In: 2017 IEEE 85th Vehicular Technology Conference (VTC Spring), pp. 1–5 (2017). https://doi.org/10.1109/VTCSpring.2017.8108218
4. Nikoukar, A., Shah, Y., Memariani, A., Gunes, M., Dezfouli, B.: Predictive interference management for wireless channels in the Internet of Things. In: 2020 IEEE 31st Annual International Symposium on Personal, Indoor and Mobile Radio Communications, pp. 1–7 (2020). https://doi.org/10.1109/PIMRC48278.2020.9217227
5. Palit, R.: Anatomy of WiFi access traffic of smartphones and implications for energy saving techniques. Int. J. Energy Inf. Commun. 3(1), 1–16 (2012)
6. Wang, D., Yang, Z.: An advanced scheme with decision tree for the improvement of spectrum sensing efficiency in dynamic network. In: 2016 9th International Congress on Image and Signal Processing, BioMedical Engineering and Informatics (CISP-BMEI), pp. 1288–1292 (2016). https://doi.org/10.1109/CISP-BMEI.2016.7852914

# Towards Road Profiling with Cooperative Intelligent Transport Systems

Mohamed-Lamine Benzagouta[1], Emilien Bourdy[2], Hasnaa Aniss[1],
Hacène Fouchal[2(✉)], and Nour-Eddin El Faouzi[1]

[1] Université Gustave Eiffel, Versailles, France
[2] Université de Reims Champagne-Ardenne, Champs-sur-Marne, France
hacene.fouchal@univ-reims.fr

**Abstract.** Cooperative Intelligent Transport Systems (C-ITS) used to generate large amounts of data from communications achieved through V2V (vehicle-to-vehicle) or V2I (vehicle-to-infrastructure). We have collected these data from a dedicated smartphone application and we have analyzed them. Useful information about road profiles and driver profiles have been provided. Various works have been dedicated to driver profiles since a decade but road profile was not of interest. We propose in this article to analyse data contained in the Cooperatives Awareness Messages (CAM) coming from anonymous participants to define road profiles considering various parameters like speed, location, day-time. The trajectories of individual vehicles are classified into various classes using four different algorithms: K-means, Agglomerative Clustering, DBSCAN and BIRCH. We have worked on 609 different trajectories and found different classes for each algorithm. Our contribution is to analyze the correlations between the different classes provided by each algorithm.

**Keywords:** C-ITS · traffic light · Connected Vehicles · Data Analysis

## 1 Introduction

Internet-of-Things (IoT) automates the execution of data-driven intelligent actions on connected devices by leveraging essential technologies such as sensor-based autonomous data collection and cloud-based big data analysis. This automation enables a wide range of practical real-world applications, including smart transportation, smart agriculture, smart cities, and so on. The incorporation of IoT into everyday life, has resulted in the collection of massive amounts of data. On Cooperative Intelligent Transport Systems (C-ITS), communications between devices is the main source of data provision. Their analysis could reveal important information about driver behaviors or road behavior. Indeed, roads could be profiled in order to detect abnormal situations. This detection opens the door for a new paradigm: real time detection without any specific device deployment. The collected data are computed by a remote server and allow to raise an alarm if a trouble occurs or if the unusual situation happens.

© The Author(s), under exclusive license to Springer Nature Switzerland AG 2024
E. Renault et al. (Eds.): MLN 2023, LNCS 14525, pp. 148–166, 2024.
https://doi.org/10.1007/978-3-031-59933-0_11

C-ITS refers to a set of technologies and systems that enable communication and cooperation between different actors of traffic network, including vehicles, infrastructure, and traffic management systems. The goal of the C-ITS is to improve safety, efficiency, and sustainability of transportation systems. It achieves this through the medium of wireless communications between the actors using several communication schemes including V2V communication, V2I communication, or vehicle-to-everything (V2X) communication. The objectives of C-ITS can be summarized as:

- Improved Safety: C-ITS can help prevent accidents by providing real-time information about road conditions, traffic, and potential hazards to drivers.
- Traffic Management: it allows for better traffic flow management by providing information to optimize traffic signal timings and control.
- Reduced emissions: C-ITS can help reduce emissions by enabling more efficient driving behaviors and traffic flow.
- Enhanced transportation efficiency: it can provide drivers with information on alternate routes, congestion, and available parking spaces, leading to smoother and faster journeys.
- Support for Autonomous and Connected Vehicles: it plays a crucial role in the development of smart and connected transportation systems, providing the communication infrastructure necessary for autonomous and semi-autonomous vehicles to operate safely and efficiently.

C-ITS plays a crucial role in the development of smart and connected transportation systems, and it is seen as a key component of future mobility solutions, including those involving autonomous vehicles. Some of its applications include collision avoidance, traffic signal management, Traffic information and Warnings, and Emergency vehicle preemption, ... and others.

Intelligent Transport Systems - G5 (ITS-G5) and Cellular Vehicle-to-Everything (C-V2X) are two different communication standards used in C-ITS. They enable the communication between the transportation actors. ITS-G5 is based on the IEEE 802.11p standard, which is an extension of the Wi-Fi standard specifically designed for vehicular communication. It operates in the 5.9 GHz spectrum band, which is allocated for transportation-related communication. Whereas C-V2X utilizes cellular networks for communication, leveraging the 4G LTE and 5G networks.

ITS-G5 is designed for short to medium-range communication, typically up to a few hundred meters, and is well suited for communication in dense urban environments and traffic scenarios. Whereas the C-V2X can provide both short and long range communications. ITS-G5 is standardized by the European Telecommunications Standards Institute (ETSI) whereas the C-V2X is standardized by the 3GPP.

Our main contribution is to handle a set of various trajectories collected from different road segments at different moments at different days. Then we split the data-set into various clusters using four different algorithms. Each algorithm gives different clusters containing different trajectories. We compare then each obtained result (from one algorithm) to all others. The clusters provided by an algorithm are different from the clusters provided by all other algorithms. We analyse these differences and what kind of information could be extracted from these differences. The paper is structured as follows. Section 2 presents some related works. Section 3 describes the well known algorithms used in data classification. Section 4 describes the obtained results and gives a general analysis of the obtained results. Finally, Sect. 5 concludes the study and gives some hints about the future directions.

## 2   Related Works

In [1] an algorithm that classifies driving style is proposed. It utilizes the statistical information from jerk profiles to classify the driving into three categories. The first one is aggressive driving with high alternating jerk, the second one is the calm driving with smooth jerk profiles, and the third one is normal driving which has moderate braking and acceleration. In [2] a framework and methodology for developing composite driver risk profiles is proposed. It uses empirical data from GPS data collected from 106 drivers in Sydney.

In [3] k-means and hierarchical clustering algorithms were run on a data set of 70 samples from a driving simulator containing number of left and right turns, number of left and right indicators, number of brakes, horns and gear change with speed. Results show that k-means performs better at classifying the data the way the authors intended, meaning three clusters slow, normal, and fast driving. In [4] GPS data (around 373 million records and 40 821 calculated speed profiles) of Taxis in the city of Zagreb was used and fed into G-means (Gaussian means) algorithm which is based on k-means. 770 clusters were determined.

In [5] smartphone sensor data was used, the data was obtained from IEEE Data port and was used to analyse driver behavior. It was clustered using the k-means method and 3 clusters were obtained were cluster 1 corresponds to aggressive/risky behavior, cluster 2 corresponds to normal driving and cluster 3 corresponds to calm driving. In [6] 4 different driver profiles were detected; safe, very safe, aggressive, and very aggressive. The profiling was done with data obtained from smartphones where a driver safety index was used to classify the profiles.

In [7] data from diverse motion sensors, including the accelerometer, gyroscope, and magnetometer, undergoes noise elimination using Kalman filter. It was then used to identify various driving events by applying time window for data extraction. The outcome of this comparison enables the classification of behavior into aggressive and non-aggressive driving profiles. In [8] they used four different machine learning algorithm; Artificial neural networks, Support Vector Machine, Random Forest, and Bayesian network on different smartphone sensor data where they identify 7 driving events; Aggressive breaking, Aggressive acceleration, Aggressive left turn, Aggressive right turn, Aggressive left lane change, Aggressive right lane change, and non-aggressive event.

In [9] the authors study the variable of vehicle Heading around some points of interest in the city of Reims. The study describes the variable in terms of min, max, mean, and median, and compares it to the heading of the point of interest. Some trajectories were extracted and clustered using different clustering approaches, the purity index of each clustering method is calculated and only the DBScan showed low scores.

The study in [10] utilized a driving simulator involving approximately 45 drivers with diverse characteristics such as gender, age, and driving experience. The simulation focused on driving toward and beyond a specified point of interest, in this case, a tunnel. The tunnel's position is regarded as the distance origin, and the starting point for recording data towards the distance origin is termed the distance length, which remains consistent after passing the point of interest. The distance length is subdivided into equivalent segments known as distance gaps, each serving as a research unit. The authors employed these distance gaps to construct a driving ethogram, from which various characteristics related to statistics on longitudinal speed, acceleration, lateral movement, and deviations from the center lanes were derived. Numerous observations and remarks were documented at the conclusion of the paper.

In [11] the focus is on extracting driving profiles from real log data collected near 22 Points of Interest (POIs). The main challenge addressed is the presence of incomplete data, where some vehicles recorded data for only one POI. Various data completion approaches, including interpolation, similar case analysis, and Bayesian methods, were employed to fill the missing data. The complete dataset was utilized for training and clustering, and it was also used to impute missing data across different ratios. The chosen ratio for analysis was 60% missing data and 40% complete data.

The work [12] examines a segment of Route National N118 in France, focusing on vehicles equipped with On-Board Units (OBUs) traveling in both directions. Data from these vehicles was extracted and analyzed using statistical methods, with a particular emphasis on speed profiles. The study provides insights into the speed profiles of Cooperative Intelligent Transport Systems (C-ITS) vehicles within a specific road segment.

Smartphone sensor data are utilized in [13], including gyroscope, accelerometer, and magnetometer readings, to distinguish between safe and unsafe driving behaviors. The sensor data undergoes preprocessing with filters to smooth acceleration and speed signals. A maneuver detection algorithm, specifically the EndPoint detection algorithm, is applied. Dynamic time warping is employed for calculating distances, and the processed signal is then subjected to a Bayesian classifier for the classification of driving instances as safe or unsafe.

## 3    Clustering Algorithms

In this section, we give the definition of the four algorithms used in this study.

### 3.1    K-Means

K-means [14] is a non supervised machine learning algorithm of non hierarchic clustering. It permits the clustering of $n$ given points in $K$ given clusters. Thus, similar observations end up in the same cluster so as to minimize a given function.

Given a set of $n$ observations $(x_1, x_2, x_3, ..., x_n)$, k-means aims to partition the $n$ observations into $k(k \leq n)$ set $S = \{s_1, s_2, ..., s_k\}$ by minimizing the distance the within-cluster sum of squares (WCSS):

$$\arg\min_s \sum_{i=1}^{k} \sum_{x_j \in s_i} \|x_j - \mu_i\|^2$$

where $\mu_i$ is the centroid of the cluster $s_i$.

For the k-means, $k$ centroids are randomly chosen from the $n$ observations at first, each centroid forms a cluster and then each observation is attributed to the nearest (mostly euclidean distance) centroid/cluster. Then the centroids of clusters are recalculated based on the mean of all the data points assigned to each cluster. The $n$ observations are assigned again to the nearest centroids and the process is repeated until convergence. Convergence occurs when the centroids no longer change significantly, indicating that the algorithm has found a stable solution.

### 3.2  Agglomerative Clustering

Agglomerative Clustering or Agglomerative Nesting (AGNESS) is a type of hierarchical clustering. The term "agglomerative" refers to the process of progressively merging or "agglomerating" groups of data points. Agglomerative clustering is a bottom-up approach, starting with individual data points and building up to larger clusters. It is in contrast to divisive clustering, which is a top-down approach that starts with all data points in one cluster and splits them into smaller clusters.

First, each observation is considered as a single element cluster, so for $n$ observations we have $n$ clusters. Second, each clusters that are close to each other are merged into one cluster, this process is repeated until one cluster remains containing all the $n$ points. Third, a dendrogram is obtained which is like a tree diagram that shows the sequence in which the clusters were merged. Last a number of clusters $k$ is chosen and the dendrogram is cut in the middle so it contains the required number of clusters.

The distance between the clusters can be defined using several methods. Some common methods include single-linkage (minimum pairwise distance), complete-linkage (maximum pairwise distance), average-linkage (average pairwise distance), and Ward's method (minimize the increase in variance within clusters).

### 3.3  BIRCH Clustering

Balanced Iterative Reducing and Clustering using Hierarchies (BIRCH) is a hierarchical clustering algorithm for large datasets efficiently [15], it works in an online incremental manner allowing it to handle datasets that might not fit completely into memory. BIRCH maintains compact summary of data using Clustering Feature (CF), CFs contain statistical information about subsets of data points. CFs are organized in a hierarchical data structure called the CF Tree. The algorithm incrementally inserts data points into the tree, selecting the appropriate leaf nodes based on criteria such as minimizing the increase in sub-cluster diameter. If inserting a point would exceed a specified diameter threshold, a new leaf node is created. As the process continues, clusters are merged and the tree is rebalanced.

### 3.4  DBSCAN Clustering

Density-Based Spatial Clustering of Applications with Noise (DBSCAN) is a density-based clustering algorithm. It does not require the number of clusters to be specified beforehand. Instead, it identifies clusters based on the density of data points in the feature space.

For $n$ observations, they can be classified as:

– Core Points: A data point is considered a core point if it has at least a specified number of neighboring points (minPts) within a certain radius ($\epsilon$).
– Border Points: A data point is considered a border point if it is within the $\epsilon - radius$ of a core point but does not have enough neighboring points to be considered a core point itself.
– Noise Points (Outliers): Data points that are neither core points nor border points are considered noise points or outliers.

The DBSCAN algorithm is given three parameters, the data set, $\epsilon$ the distance which makes a cluster and $minPts$ as the minimum points that can constitute a cluster. It then selects randomly points that were not visited before, If the selected point has at least $minPts$ neighbours with $\epsilon$ distance, then it is considered a core point and the same process is done for its neighbouring points and a cluster is created with the core and border points that were visited.

## 4  Trajectory Analysis

CRoads is one of the joint initiative projects of European member states that aim to developing innovative C-ITS solutions. The architecture of CRoads is very rich and comprises of both long and short range communications. As part of the project, the CoopITS application was developed, in which a smartphone plays the role of a station (a vehicle). It allows the sending and reception of C-ITS messages using the cellular network. The application was launched in January 2021 and it functions mainly in the region of "La Nouvelle-Aquitaine" in France.

The application records all C-ITS messages which get stored, such as CAMs, DENMs, SPATEMs, MAPEMs, and IVIMs, which are recorded in their raw ASN.1 UPER encoded state, and then are decoded and stored into a data base.

We are mainly interested in the CAM messages which are periodically sent by each C-ITS station. The frequency of the CAM message in C-ITS standards is of between 100 ms and 1 s depending on the velocity, but its frequency in the CoopITS application is either 1 s or 5 s.

A CAM message contains information about the station's state in an instant $t$, such as its velocity, heading and position in means of longitude and latitude. Each station has a unique pseudonym called the stationID which is attributed randomly and gets updated every 10 min for anonymity.

Having the set of CAM messages, we wanted to find a specific pattern, which is a trajectory that was traversed several times, and from the dataset containing 1 336 573 CAMs coming from 8 667 distinct stationIDs from the region "Nouvelle-Aquitaine". At the time of the study we identified a single trajectory that was traversed 71 times (71 distinct stationIDs) in the same direction in the city of Bordeaux which contains a total of 46 570 CAM messages.

Figure 1 shows the trajectory that was chosen. The trajectory starts south and goes to the north and was plotted on an OSM map in python using the folium library where each dot represents the position recorded in each CAM message from a single trip (a single stationID).

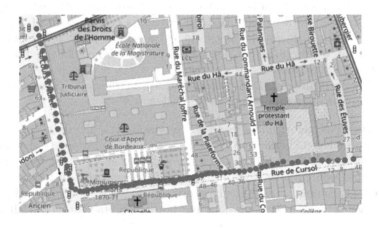

**Fig. 1.** An example of the trajectory that was extracted from CAM messages, plotted on top of an OSM map

We then cut the trajectory into 10 elementary road segments, a segment is the portion of road between two successive road crossings. Figure 2 shows the segments that were cut from the trajectory.

To determine which CAM point (latitude and longitude) belong to which segment, we draw a rectangle out of the segment, its length is the length of the segment and its width is of 20 m, if the point belongs to the rectangle it gets attributed to it. We chose a large width due to GPS precision issues, as a large portion of the CAM points were out of the borders of the road.

After doing so, we ended up with 609 distinct trajectories, 71 multiplied by 10 gives 710 which should be the number of trajectories to end up with, but we had 609 and this is due to the large frequency of the CAM messages, certain road segment are small and don't fit to a single CAM point.

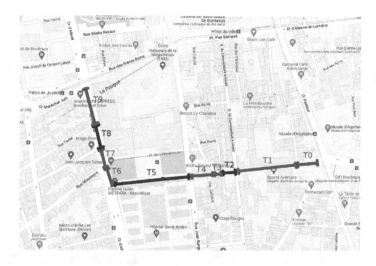

**Fig. 2.** The determined segments, plotted on Google maps

609 different trajectories have been collected. They have used as an input for each of the four algorithms.

The obtained trajectories have different dimensions so we can't feed them into clustering algorithms which require that all observations have the same dimensions. So we converted each trajectory into a 60 dimension observation by projecting 60 points on the curve of velocity of each trajectory.

## 4.1   K-Means Analysis

We have run the k-means algorithm on the trajectory data. We have chosen 5 clusters, Table 1 shows the number of classification per cluster, the first cluster is the largest as it has 184 observations. The cluster 4 contains the less number of observations. Figure 3 shows samples of trajectories of the different clusters. We can see from the samples that cluster 0 and cluster 1 are similar in their shapes which is a free flow without a stop with a velocity lower than 6 m/s. Whereas cluster 4 is a free flow with velocity that is higher than 8 m/s. Clusters 2 and 3 both have trajectories with a stop (with a traffic light or a queue) but they differ by the time of the stop with the cluster 3 having the largest time of stop.

**Table 1.** Kmeans: number of trips for each cluster

| cluster | nbrTrips |
|---------|----------|
| 0       | 184      |
| 1       | 109      |
| 2       | 94       |
| 3       | 136      |
| 4       | 86       |

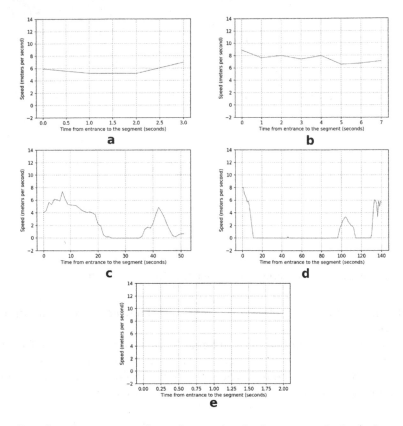

**Fig. 3.** Samples of trajectories from clusters for the kmeans method: a) cluster 0, b) cluster 1, c) cluster 2, d) cluster 3, e) cluster 4.

## 4.2  Agglomerative Clustering Analysis

We have run the Agglomerative Clustering algorithm on the trajectory data and chose 5 clusters. As can be seen in Table 2 cluster 0 has the largest number of samples with 219 observations, and with cluster 3 having the lowest number of trajectories. As can be seen in Fig. 4 clusters 3 and 0 are free flow trajectories which are lower than 9 m/s, whereas cluster 1 represents free flow trajectories that are relatively higher than 8 m/s. Clusters 2 and 4 represent trajectories with a stop with cluster 2 having smaller stopping intervals.

**Table 2.** Agglomerative clustering: number of trips for each cluster

| cluster | nbrTrips |
|---------|----------|
| 0 | 219 |
| 1 | 111 |
| 2 | 85 |
| 3 | 92 |
| 4 | 102 |

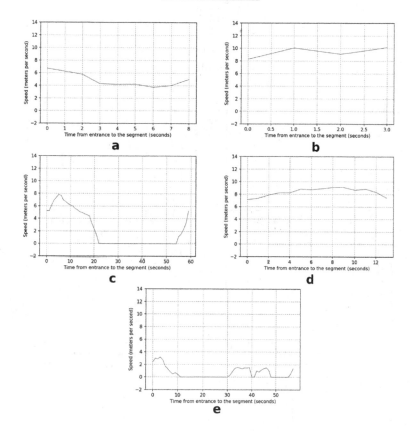

**Fig. 4.** Samples of trajectories from clusters for the Agglomerative clustering method: a) cluster 0, b) cluster 1, c) cluster 2, d) cluster 3, e) cluster 4.

## 4.3 BIRCH Clustering Analysis

We have run the BIRCH algorithm and chose 5 clusters. As can be seen in Table 3 cluster 0 has the largest number of observations and cluster 3 has the lowest number of observations. As can be seen in Fig. 5 clusters 0 and 1 represent the free flow trajectories that have velocities relatively lower than 9 m/s. Whereas cluster 3 represents the free flow trajectories with velocities that are relatively larger than 9 m/s. Cluster 2 and 4 have trajectories with a stop in the middle of the segment.

**Table 3.** BIRCH: number of trips for each cluster

| cluster | nbrTrips |
|---------|----------|
| 0 | 219 |
| 1 | 142 |
| 2 | 85 |
| 3 | 61 |
| 4 | 102 |

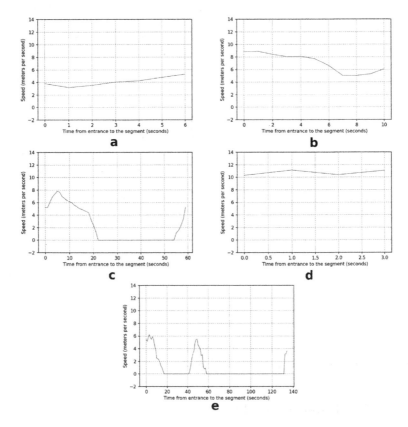

**Fig. 5.** Samples of trajectories from clusters for the BIRCH clustering method: a) cluster 0, b) cluster 1, c) cluster 2, d) cluster 3, e) cluster 4.

## 4.4   DBSCAN Clustering Analysis

We have run the DBSCAN algorithm and we have obtained 6 different clusters. The cluster 0 contains 81% of the observations, and 15% of the observations are considered aberrant. Cluster 1 contains 7 observations and clusters 2, 3, 4, and 5 contain each 3 observations. The reason the cluster 0 contains 81% of the observations is that the dataset is continually dense as can be seen in Fig. 6 (Table 4 and Fig. 7).

**Table 4.** DBSCAN: number of trips for each cluster

| cluster | nbrTrips |
| --- | --- |
| −1 | 95 |
| 0 | 495 |
| 1 | 7 |
| 2 | 3 |
| 3 | 3 |
| 4 | 3 |
| 5 | 3 |

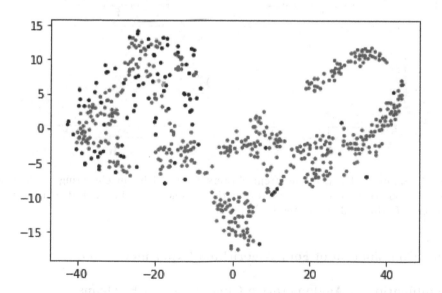

**Fig. 6.** DBSCAN clusters visualized using T-SNE

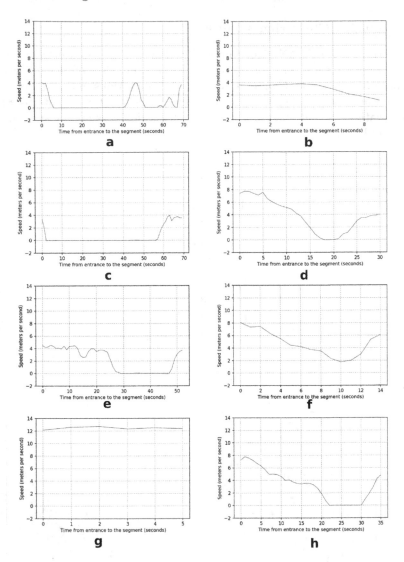

**Fig. 7.** Samples of trajectories from clusters for the Dbscan clustering method: a) aberrant observation, b) cluster 0, c) anothe sample from cluster 0, d) cluster 1, e) cluster 2, f) cluster 3, g) cluster 4, h) cluster 5.

## 4.5 Combination of Four Algorithms Experimentation

### Combination of Agglomerative Clustering and K-Means

Table 5 shows a comparison between the AC and K-means, we notice that 175 observations are in clusters 0 and 0 in AC and K-means, cluster 0 in k-means has 8 and 1 observations common with clusters 3 and 4, which allows us to safely say that clusters 0, and 0 in the AC and k-means are relatively equivalent which represent the free flow with velocities generally lower than 6 m/s. The same can be observed in regard to cluster 3 from the k-means, as it has 101 common observations with cluster 4 of the AC, and 18, 17 common observations with clusters 2 and 0 of the AC respectively, thus we can assume that clusters 4 and 3 of the AC and k-means respectively relatively constitute the same cluster.

From a one sided perspective, we notice that cluster 4 of the k-means has 85 common observations with cluster 1 of the AC and has one common observation with cluster 3 of the AC. So we can assume that cluster 1 mostly includes cluster 4.

**Table 5.** Comparison between the AC and k-means

| AC = Kmeans | number |
|-------------|--------|
| 0 = 0       | 175    |
| 0 = 2       | 27     |
| 0 = 3       | 17     |
| 1 = 1       | 26     |
| 1 = 4       | 85     |
| 2 = 2       | 67     |
| 2 = 3       | 18     |
| 3 = 1       | 83     |
| 3 = 0       | 8      |
| 3 = 4       | 1      |
| 4 = 3       | 101    |
| 4 = 0       | 1      |

## Combination of BIRCH and Agglomerative Clustering

In Table 6 a comparison between the BIRCH and AC algorithms. We observe that clusters 0 and 0 of the BIRCH and AC algorithms are equivalent and are in a single instance, thus it means that the clusters are the same. We observe the same thing with clusters 2 and 4 of both BIRCH and AC with 85 and 102 common observations respectively, meaning that cluster 0 is equal to 0 from the AC, cluster 2 is equal to 2 of the AC, and cluster 4 is equal to 4 from the AC. The resemblance between the results of both algorithms is because they're both hierarchical algorithms.

**Table 6.** Comparison between the BIRCH and AC

| BIRCH = AC | number |
|---|---|
| 0 = 0 | 219 |
| 1 = 1 | 50 |
| 1 = 3 | 92 |
| 2 = 2 | 85 |
| 3 = 1 | 61 |
| 4 = 4 | 102 |

## Combination of BIRCH and Dbscan

In Table 7 we see a comparison between the algorithms BIRCH and DBSCAN, and we see that cluster 0 of the DBSCAN is distributed between clusters 0, 1, 2, 3, and 4 of the BIRCH. Clusters 1, 3, and 5 of the DBSCAN are included in cluster 0, cluster 2 of DBSCAN is included in cluster 2 of the BIRCH and cluster 4 of DBSCAN is included in cluster 3 of the BIRCH.

**Table 7.** Comparison between the BIRCH and DBSCAN

| BIRCH = DBSCAN | number |
|---|---|
| 0 = 0 | 180 |
| 0 = -1 | 26 |
| 0 = 5 | 3 |
| 0 = 3 | 3 |
| 0 = 1 | 7 |
| 1 = 0 | 138 |
| 1 = -1 | 4 |
| 2 = -1 | 36 |
| 2 = 0 | 46 |
| 2 = 2 | 3 |
| 3 = 0 | 57 |
| 3 = 4 | 3 |
| 3 = -1 | 1 |
| 4 = -1 | 28 |
| 4 = 0 | 74 |

## Combination of BIRCH and K-Means

In Table 8 the comparison between BIRCH and K-means results. WE observe that cluster 0 of the K-means and cluster 0 of BIRCH have 175 observations in common. Cluster 0 of k-means and cluster 1 of BIRCH have 8 common observations and also that cluster 0 of k-means and cluster 4 of BIRCH have one observation in common, which means that cluster 0 is almost included in cluster 0 of BIRCH. We also observe that cluster 1 of k-means is included in cluster 1 of BIRCH and have 109 common observations. Cluster 3 from k-means is almost included in cluster 4 of BIRCH, with 17 observations in cluster 0 of the BIRCH algorithm.

**Table 8.** Comparison between the BIRCH and K-means

| BIRCH = K-means | number |
|---|---|
| 0 = 0 | 175 |
| 0 = 2 | 27 |
| 0 = 3 | 17 |
| 1 = 1 | 109 |
| 1 = 4 | 25 |
| 1 = 0 | 8 |
| 2 = 2 | 67 |
| 2 = 3 | 18 |
| 3 = 4 | 61 |
| 4 = 3 | 101 |
| 4 = 0 | 1 |

## Combination of Dbscan and Agglomerative Clustering

In Table 9 a comparison between the algorithms DBSCAN and AC and we see that cluster 0 of the DBSCAN is distributed between clusters 0, 1, 2, 3, and 4 of the AC. Clusters 1, 2, 3, 4, and 5 of DBSCAN are distributed between (and included in) 0, 2, and 1 of the AC.

**Table 9.** Comparison between the DBSCAN and AC

| dbscan = AC | number |
|---|---|
| −1 = 2 | 36 |
| −1 = 4 | 28 |
| −1 = 0 | 26 |
| −1 = 1 | 3 |
| −1 = 3 | 2 |
| 0 = 1 | 105 |
| 0 = 0 | 180 |
| 0 = 3 | 90 |
| 0 = 4 | 74 |
| 0 = 2 | 46 |
| 1 = 0 | 7 |
| 2 = 2 | 3 |
| 3 = 0 | 3 |
| 4 = 1 | 3 |
| 5 = 0 | 3 |

## Combination of Dbscan and K-Means

In Table 10 a comparison between DBSCAN and K-means. And we see that cluster 0 of the DBSCAN is distributed between clusters 0, 1, 2, 3, and 4 of the K-means. Clusters 1, 2, 3, and 4 of DBSCAN are distributed between clusters 0, 2, and 4 of the K-means

**Table 10.** Comparison between the DBSCAN and K-means

| dbscan = Kmeans | number |
|---|---|
| −1 = 2 | 36 |
| −1 = 3 | 41 |
| −1 = 0 | 13 |
| −1 = 1 | 4 |
| −1 = 4 | 1 |
| 0 = 1 | 105 |
| 0 = 0 | 168 |
| 0 = 4 | 82 |
| 0 = 3 | 95 |
| 0 = 2 | 45 |
| 1 = 2 | 7 |
| 2 = 2 | 3 |
| 3 = 0 | 3 |
| 4 = 4 | 3 |
| 5 = 2 | 3 |

# 5   Conclusion

We have shown in this paper that connected vehicles are an important source of data provision. This issue could be exploited positively to solve some troubles as abnormal situation detection without any additional device on roads. Indeed, in this study we have collected various trajectories on various roads at different periods. We aimed to classify all these trajectories into different clusters. Then each cluster has to be checked in order to understand what are the features of each cluster. We have applied 4 clustering algorithms in order to check the accuracy of each of them.

The main conclusion is that each algorithm has its own classification and the intersections between clusters from one algorithm to another is not empty. Afterwards, we have focused on the differences of the obtained results by each algorithm. These differences are of high interest since they express a king of sub-classes having their own features.

# References

1. Murphey, Y.L., Milton, R., Kiliaris, L.: Driver's style classification using jerk analysis. In: 2009 IEEE Workshop on Computational Intelligence in Vehicles and Vehicular Systems, pp. 23–28 (2009)
2. Ellison, A.B., Greaves, S.P., Daniels, R.: Profiling drivers' risky behaviour towards all road users (2012)
3. Kalsoom, R., Halim, Z.: Clustering the driving features based on data streams. In: INMIC, pp. 89–94 (2013)
4. Erdelić, T., Vrbančić, S., Rošić, L.: A model of speed profiles for urban road networks using G-means clustering. In: 2015 38th International Convention on Information and Communication Technology, Electronics and Microelectronics (MIPRO), pp. 1081–1086 (2015)
5. Anil, A.R., Anudev, J.: Driver behavior analysis using K-means algorithm. In: 2022 Third International Conference on Intelligent Computing Instrumentation and Control Technologies (ICICICT), pp. 1555–1559 (2022)
6. Saiprasert, C., Thajchayapong, S., Pholprasit, T., Tanprasert, C.: Driver behaviour profiling using smartphone sensory data in a V2I environment. In: 2014 International Conference on Connected Vehicles and Expo (ICCVE), pp. 552–557 (2014)
7. Wu, M., Zhang, S., Dong, Y.: A novel model-based driving behavior recognition system using motion sensors. Sensors **16**, 1746 (2016)
8. Ferreira, J., et al.: Driver behavior profiling: an investigation with different smartphone sensors and machine learning. PLoS ONE **12**, 1–16 (2017)
9. Leblanc, B., Fouchal, H., de Runz, C.: Driver profile detection using points of interest neighbourhood. In: 2019 IEEE 90th Vehicular Technology Conference (VTC2019-Fall), pp. 1–4 (2019)
10. Mao, Y., Zhang, W., Wang, M., Guo, D.: Driving simulator data based driver behavior ethogram establishment. In: 2019 1st International Conference on Industrial Artificial Intelligence (IAI), pp. 1–5 (2019)
11. Leblanc, B., Ercan, S., de Runz, C.: C-its data completion to improve unsupervised driving profile detection. In: 2020 IEEE 91st Vehicular Technology Conference (VTC2020-Spring), pp. 1–5 (2020)

12. Moso, J.C., Cormier, S., Fouchal, H., de Runz, C., Wandeto, J.M., Aniss, H.: Road speed signatures from C-its messages. In: ICC 2021 - IEEE International Conference on Communications, pp. 1–6 (2021)
13. Eren, H., Makinist, S., Akin, E., Yilmaz, A.: Estimating driving behavior by a smartphone. In: 2012 IEEE Intelligent Vehicles Symposium, pp. 234–239 (2012)
14. MacQueen, J.: Some methods for classification and analysis of multivariate observations (1967)
15. Zhang, T., Ramakrishnan, R., Livny, M.: BIRCH: an efficient data clustering method for very large databases. SIGMOD Rec. **25**, 103–114 (1996)

# Study of Masquerade Attack in VANETs with Machine Learning

Yasmine Chaouche[1], Éric Renault[2(✉)], and Ryma Boussaha[1]

[1] École nationale Supérieure d'Informatique Algiers, Oued Smar, Algeria
{hy_chaouche,r_boussaha}@esi.dz
[2] ESIEE Paris, Paris, France
eric.renault@esiee.fr

**Abstract.** Vehicular Ad Hoc Network (VANET) affords communication between vehicles and roadside infrastructures in order to improve road safety and driving conditions. The implementation of precise security mechanisms is imperative in VANET to ensure safety, as attackers are constantly seeking ways to exploit network vulnerabilities. Among the prevalent risks, the masquerade attack stands out as a common and impactful threat. Masquerade is performed by malicious nodes to gain access to the network by impersonating the real identity of a legitimate vehicle. In this paper, we evaluate three variants of masquerade with different machine learning algorithms. These experiments are realised using the F2MD simulation environment with Weka. The performance of the misbehavior detection mechanisms is evaluated by accuracy, recall, and precision. Our comparative results demonstrate that the Random Forest (RF) algorithm outperforms the others in the proposed simulation scenario in terms of accuracy, while the decision tree algorithm (J48) excels as the fastest with minimal prediction time.

**Keywords:** VANET · Masquerade attack · F2MD · Machine Learning

## 1 Introduction

Vehicular Ad-Hoc Networks (VANETs) have appeared as a subset of Mobile Ad-Hoc Networks (MANETs) [1] and are considered one of the most promising fields in wireless networks. There are basically two types of communication in VANETs: Vehicle-to-Vehicle (V2V) and Vehicle-to-Infrastructure (V2I). These communications enable vehicles to share information such as location and accident warnings. VANETs face various challenges and security issues that researchers need to address. Due to the special characteristics of VANET networks, such as their high mobility, they are vulnerable to various types of internal and external attack. In fact, the whole VANET communication system is an open-access environment, which allows attackers to modify, inject and delete messages. For instance, attackers could gain access to traffic messages used to guide vehicles on the road. By altering these messages, they can spread false information about the road conditions, leading to potential dangers.

É. Renault et al. (Eds.): MLN 2023, LNCS 14525, pp. 167–184, 2024.
https://doi.org/10.1007/978-3-031-59933-0_12

One common and harmful threat is the masquerade attack [8], also known as identity theft. This attack occurs when a user does not reveal their true identity (using a valid mask to conceal themselves) and pretends to be another user in order to gain unauthorized access, receive messages, or obtain privileges not granted to them. One example of using this type of attack is when an attacker pretends to be an emergency vehicle, compelling other vehicles on the road to change lanes or reduce their speed [6]. To ensure the security of VANETs, different security measures are designed, with the most popular being IDS (Intrusion Detection System) [12]. While IDS has proven effective in detecting malicious nodes in traditional networks, its application in VANETs is somewhat different and challenging due to their unique characteristics. Therefore, machine learning has become one of the most widely used techniques in cybersecurity, especially for IDS [10]. These intelligent models offer greater flexibility, enabling them to detect potential threats and provide effective solutions.

In this work, we propose a study on masquerading attacks using machine learning algorithms, which, to our knowledge, have not been explored for vehicular networks in the existing literature. To this end, we employed the Framework for Misbehavior Detection (F2MD) [3,9] and the Luxembourg scenario [4] to evaluate three types of masquerade attacks: Data Replay Masquerade, DoS Disruptive Masquerade, and DoS Random Masquerade. We first injected and then evaluated these attacks using different ML-based misbehavior detection methods using the WEKA tool (Waikato Environment for Knowledge Analysis) [5]. Our comparative results show that the RF model outperforms the others with greater accuracy, and the J48 model is the fastest in terms of prediction time in our proposed scenario.

The rest of the paper is organized as follows: Sect. 2 describes the algorithms and schemes we propose for the three types of masquerade attacks and explains how they can be presented in the VANET system. In Sect. 3, we introduce the simulation settings and scenarios used for data collection. Section 4 presents the results and analysis of the performance comparison for our machine learning algorithms. Finally, Sect. 5 concludes the paper and presents future work.

## 2   Masquerade Attack

The masquerade attack [6,8], also known as identity spoofing attack, involves the use of a fake identity by the attacker to act as another entity. This attack occurs when a user, instead of revealing his true identity, pretends to be someone else. To do this, he uses a valid mask (his pseudonym, which is the same as the victim's) to hide his true identity and impersonate another entity in order to gain unauthorized access, receive messages or obtain privileges that are not legitimately accorded to him.

Masquerade attacks come in several variants depending on the attacker's objectives and level of knowledge. The motivation behind such attacks may include:

- Overwhelming the behavior detection algorithms of neighbouring stations in the intelligent transportation system.
- Degrading the quality of the security system, thereby reducing the reliability of information exchanges.
- Testing the detection effectiveness by employing a combination of multiple attacks (increasing the complexity of classification for example).

To compare and analyze different masquerade attacks, we refine the definition, providing a more in-depth description and explaining how these attacks impact the VANET network. Through this work, we propose three distinct categories: DoS Random Masquerade, Data Replay Masquerade, and DoS Disruptive Masquerade. Each type represents a combination of masquerade and one or two other attack types (DoS Random, Data Replay and DoS Disruptive). We present below the detailed algorithm we propose, as well as a descriptive overview of each of these three types of attack.

## 2.1  DoS Random Masquerade

The objective of a DoS (Denial of Service) [7] attack is to prevent legitimate users from receiving a message. To perform this type of attack, the malicious vehicle sends V2X (vehicle-to-vehicle or vehicle-to-infrastructure communication) [11] messages at a frequency higher than that defined in the standard. This increase in frequency overloads the broadcast channel, making it unusable for other vehicles. As illustrated in Fig. 1, the attacker generates messages containing random data (e.g., a false position). In this scenario, it uses the same pseudonym as the targeted legitimate vehicle to transmit malicious messages (e.g., a request to slow down).

Algorithm 1 outlines our proposed scheme for the DoS Random Masquerade attack. The inputs include a list of real pseudonyms existing in the network, the frequency of DoS message transmission, and the maximum values for position, speed, and acceleration. First, we gather the list of real pseudonyms of vehicles on the network and randomly select one pseudonym to target for the attack. Subsequently, we initialize the DoS attack by modifying the message transmission interval. The message to be sent by the attacking vehicle is generated by assigning it the pseudonym of the selected sender vehicle. Finally, for all message characteristics such as position and speed, we assign randomly generated values within the maximum ranges of these attributes. Other values, including the length and width of the transmitting vehicle, as well as confidence values, are also assigned. The output of this algorithm is the generated message.

## 2.2  Data Replay Masquerade

In the Data Replay attack, the malicious vehicle selects a target and replays its data instantly. The attacker then generates messages containing the positions broadcasted by the legitimate vehicle. Consequently, for an observer, it would seem as if there are two vehicles existing in the same spatio-temporal dimension.

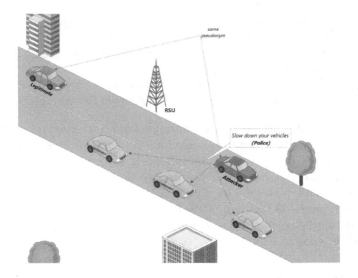

**Fig. 1.** Masquerade attack: DoS Random

Figure 2 provides an example of the application of the Data Replay Masquerade attack. At time $t = t_1$, the attacker sends a message containing the same position $X_1$ as the victim vehicle. One of the significant challenges for a detection system in this attack is the difficulty of distinguishing between the true node (the victim) and the malicious node, as they both use the same pseudonym. In this case, there is a high probability that the legitimate vehicle might be mistakenly identified as the attacker.

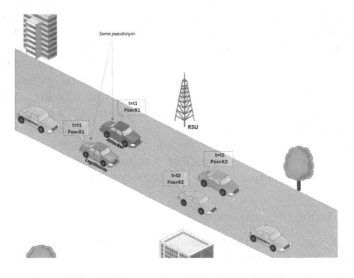

**Fig. 2.** Masquerade attack: Data Replay

---

**Algorithm 1:** DoS Random Masquerade Algorithm

---

**Inputs:** list of existing real pseudonyms (*RealPseudonymList*), frequency of
DoS message transmission (*localDoSMultipleFreq*), maximum values for
position, speed, and acceleration
**Outputs:** Malicious message (*attackBsm*)
Let *DoSInitiated* be a DoS initialization boolean
*DoSInitiated*=false
value ← RandomChoice(*RealPseudonymList*)
**if** *(DoSInitiated=false)* **then**
|   Let *beaconInterval* be the interval between two transmitted messages.
|   *beaconInterval*←(*beaconInterval* / *localDoSMultipleFreq*)
|   *DoSInitiated* = true
**end**
*attackBsm*← Generate the message to be sent
*attackBsm.Pseudonym*←value ;    /* Assign the randomly chosen pseudonym
value to the malicious message */
*attackBsm.Posistion* ← Generate random values for position coordinates
between 0 and the maximum position values
*attackBsm.PosConfidence* ← position confidence value
*attackBsm.Speed* ← Generate random values for speed coordinates between 0
and the maximum speed values.
*attackBsm.SpeedConfidence*← speed confidence value
*attackBsm.Accel*← Generate random values for acceleration coordinates
between 0 and the maximum acceleration values.
*attackBsm.AccelConfidence*← acceleration confidence value
*attackBsm.Heading*← Generate a random heading value (between [-1,1])
*attackBsm.HeadingConfidence*← heading confidence value
*attackBsm.Width* ← vehicle width
*attackBsm.Length* ← vehicle length
**return** *attackBsm*

---

The algorithm for the Data Replay Masquerade attack that we propose is
presented in Algorithm 2. As in the previous attack, we first randomly select a
vehicle to attack by using its pseudonym. To replay its data, we predefine the
number of messages to replay. The attacker resends the messages already received
by its vehicle with the stolen pseudonym, resulting in two vehicles sending the
same message (attacker and legitimate) at the same instant.

---

**Algorithm 2:** Data Replay Masquerade Algorithm

---

**Inputs:** list of existing real pseudonyms (*RealPseudonymList*)
**Outputs:** Malicious message (*attackBsm*)
value ← RandomChoice(*RealPseudonymList*)
Let *nodeHistoryList* be the list of neighboring nodes that have previously
  sent messages to the attacking vehicle.
Let *NumdetectedNode* be the number of neighboring nodes that have
  previously sent messages to the attacking vehicle.
Let *localReplaySeqNum* be the number of messages to replay instantly
  from a neighboring vehicle.
**if** *(NumdetectedNodes >0)* **then**
  **if** *(ReplaySeq <localReplaySeqNum)* **then**
    Node← the node with pseudonym Value
    attackBsm ← choose the message instantly received from the Node
      whose position is closest to that of the attacking vehicle.
    ReplaySeq ++;
  **else**
    Node← RandomChoice(*nodeHistoryList*).
    *attackBsm* ← randomly choose a message to replay from Node's
      history.
    ReplaySeq = 0;
  **end**
  *attackBsm.Pseudonym*←value ;     /* Assign the randomly chosen
    pseudonym value to the malicious message */

  **return** *attackBsm*
**end**

---

## 2.3 DoS Disruptive Masquerade

In the Disruptive attack, the malicious vehicle records data broadcasted by neighboring nodes. The attacker then disseminates V2X messages containing previously received data. The goal of the attacker in this case is to inundate the network with such messages, thereby degrading the quality of service for cooperative transportation systems.

The DoS Disruptive Masquerade attack is a combination of the two preceding attacks. As shown in Fig. 3, the attacker uses the pseudonym of a legitimate vehicle within the network to send messages without filling them with random data. Instead, the transmitted data is based on information received from nearby vehicles (using the history of received messages). Unlike the Data Replay Masquerade attack, the attacking vehicle in the DoS Disruptive Masquerade does not follow a specific victim; instead, it aims to disrupt the system with sudden appearances at separate positions.

In Fig. 3, the attacker transmits a message at time $t = t_1$, containing one position ($pos = X_1$), and at time $t = t_2$, sends another message with a dis-

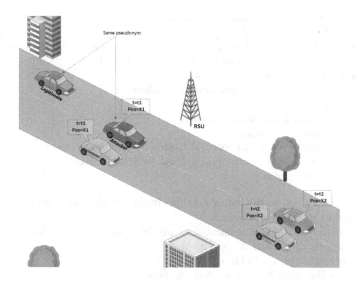

**Fig. 3.** Masquerade attack: DoS Disruptive

tinct position ($pos = X_2$), which represents the position of another vehicle. The attacker's behavior is motivated by the deterioration in the quality of the security system, which leads to a reduction in the reliability of the exchanged information.

We propose the Algorithm 3 for the DoS Disruptive Masquerade attack which combines the two previous algorithms. The input set includes the list of real pseudonyms existing in the network and the frequency of DoS message transmission. First, we randomly select a pseudonym from those existing in the network. Next, we initiate the DoS attack by modifying the interval between two transmitted messages. We randomly select a neighboring vehicle whose data we intend to replay. Next, we randomly select a message already received by the attacker from this vehicle. Finally, we assign the value of the pseudonym to the selected attacking message, which will be returned as an output of this algorithm.

## 3 Simulation Settings and Scenarios

To implement the three types of masquerade attacks with different levels of complexity, as defined in Sect. 2, we utilize our previous enhancement of the F2MD framework [3]. F2MD [9] includes various tools for the real-time simulation and evaluation of an MisBehavior Detection (MBD) system. This framework extends VEINS [13], an open-source inter-vehicular communication (IVC) simulation framework made up of a network simulator (OMNeT++) [14] and a road traffic simulator (SUMO) [2]. OMNeT++ provides all the necessary conditions to simulate various security attacks. The implementation of the masquerade attack involves randomly selecting a vehicle in the network and choosing it for the attack.

---

**Algorithm 3:** DoS Disruptive Masquerade Algorithm

---

**Inputs:** list of existing real pseudonyms (*RealPseudonymList*), frequency of
　　　　DoS message transmission (*localDoSMultipleFreq*)
**Outputs:** Malicious message (*attackBsm*)
Let *DoSInitiated* be a DoS initialization boolean.
*DoSInitiated*=false
value ← RandomChoice(*RealPseudonymList*).
**if** *(DoSInitiated=false)* **then**
　| 　Let *beaconInterval* be the interval between two transmitted messages.
　| 　*beaconInterval*←(*beaconInterval* / *localDoSMultipleFreq*)
　| 　*DoSInitiated* = true
**end**
Let *nodeHistoryList* be the list of neighboring nodes that have previously sent
　messages to the attacking vehicle.
Let *NumdetectedNode* be the number of neighboring nodes that have previously
　sent messages to the attacking vehicle.
**if** *(NumdetectedNodes >0)* **then**
　| 　Node← RandomChoice(*nodeHistoryList*).
　| 　*attackBsm* ← randomly choose a message to replay from Node's history.
　| 　*attackBsm.Pseudonym*←value ;　　　　/* Assign the randomly chosen
　| 　pseudonym value to the malicious message */
　|
　| 　**return** *attackBsm*
**end**

---

All experiments are conducted on a virtual machine running Ubuntu 18.04 with 10 GB of RAM, 4 cores, and a disk space of 89 GB. The virtual machine is hosted on VMware Workstation, which operates on a PC equipped with an Intel(R) Core(TM) i7-8650U processor @ 1.90 GHz 2.11 GHz and 16.0 GB of RAM. We obtain the final results by running simulations with different vehicle densities and using the simulation parameters as defined in Table 1.

In the DoS Random Masquerade attack, attacking vehicles are configured to send messages containing stochastically generated random data with increased frequency. For the Data Replay Masquerade attack, we replayed the data received by the attacking vehicle at the simulation time. The DoS Disruptive Masquerade attack is implemented by randomly choosing the position, speed, and acceleration of a legitimate node among the existing nodes. All masquerade parameters, as described in Table 2, are specified in the main configuration file, Omnetpp.ini. In the Veins sub-project, we have implemented masquerade attack algorithms, adapting them based on parameters associated with different scenarios, including those specified in the Table 2.

To collect our data, we use a simulation scenario for training and testing, as shown in Fig. 4. This scenario represents a portion of the original LuST scenario [4]. The size of this network is 6.51 km$^2$ with an average density of 104.5 vehicles/km$^2$. It is worth noting that two simultaneous simulations of the same scenario do not generate the same dataset. Indeed, all information broadcasted

**Table 1.** Simulation parameters

| Parameter | Value |
|---|---|
| MAC protocol | IEEE 802.11p |
| Transmission range | 420 m, 2600 m |
| Speed | Maximum road speed |
| Transmission power | 20 mW |
| Sensitivity (minPowerLevel)[a] | −89 dBm |
| Bit rate | 6 Mbps |
| Channel frequency | 5.89 GHz |
| Beacon interval | 1 s |
| Update interval[b] | 0.1 s |
| Beacon size | 256 bits |
| Node placement | Random |
| Route generation | Random |

[a] The minPowerLevel indicates the minimum reception power needed to decode a frame.
[b] Time interval used to update the position of vehicles.

**Table 2.** Parameters of masquerade attacks

| | Parameter | Value |
|---|---|---|
| DoS Random Masquerade | parVar | 0.55 |
| | MaxRandomPos | playgroundSize |
| | curPositionConfidence | random[a] |
| | RandomSpeedX | 40 |
| | RandomSpeedY | 40 |
| | curSpeedConfidence | random[b] |
| | RandomAccelX | 2 |
| | RandomAccelY | 2 |
| | curHeadingConfidence | random[c] |
| Data Replay Masquerade | ReplaySeqNum | 6 |
| | localReplaySeqNum | random[d] |
| DoS Disruptive Masquerade | DosMultipleFreq | 4 |
| | localDosMultipleFreq | random[e] |

[a] Value = Coord(Random(3, 5), Random(3, 5), 0)
[b] Value = Coord(Random(0, 0.0016), Random(0, 0.0016), 0)
[c] Value = Coord(Random(0, 20), Random(0, 20), 0)
[d] Value = Random((1-parVar), (1+parVar)) * ReplaySeqNum
[e] Value = Random((1-parVar), (1+parVar)) * DosMultipleFreq

by vehicles differs, including their positions, plausibility check values, and the number of messages sent and received by a given vehicle. Additionally, as the attacker intensity in the training scenario differs from that in the testing scenario, the number and timing of generating attacker messages (`CreationTime`) are different, and the timing of message generation or reception cannot be the same.

(a) Small LuST Scenario

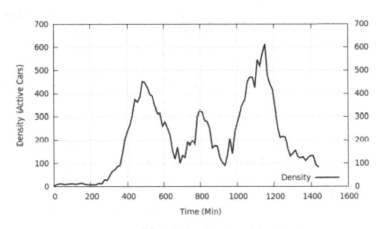

(b) Vehicle Density

**Fig. 4.** Scenario and Vehicle Density

The training dataset contains 254,499 transmitted messages (5 h of simulation). Table 3 shows the total number of normal messages and masquerade attacks in the training dataset. The real-time test scenario consists of 40,090 messages transmitted over 2.8 h. In both the training and test datasets, we randomly inject masquerade attacks with an attack rate of 25% during the training scenario and 5% for the test scenario.

**Table 3.** Occurrence of masquerade attack instances in the training dataset.

| Attack Type | Number of Instances |
|---|---|
| Genuine | 124,673 |
| DoSRandomMasquerade | 36,744 |
| DataReplayMasquerade | 42,893 |
| DoSDisruptiveMasquerade | 50,189 |

To be able to compare the results of masquerade attack detection with machine learning in VANET networks, we have chosen four algorithms from WEKA. These algorithms are commonly employed for attack detection and include J48, a variant of the C4.5 decision tree, Random Forest (RF), Naïve Bayes (NB) and Support Vector Machine (SVM) implemented using sequential minimal optimization (SMO) [15].

# 4    Results and Analysis

## 4.1    Performance Evaluation

All trained models use messages collected in real time for testing. We evaluate our experimental results using a variety of evaluation metrics, which are derived from the following four fundamental concepts that form the matrix for binary misbehavior classification: (1) True Positive (TP) denotes the accurate identification of malicious vehicles, (2) False Positive (FP) refers to the misclassification of legitimate nodes as malicious vehicles, (3) True Negative (TN) signifies the correct identification of legitimate nodes as non-malicious vehicles, and (4) False Negative (FN) represents the erroneous classification of malicious nodes as legitimate vehicles.

We evaluated classifier performance by examining various parameters, which include standard measures namely Accuracy, False Positive Rate (FPR), Recall, Precision and F1-score. In addition, we consider more specialized measures like Matthews Correlation Coefficient (MCC), Cohen's Kappa ($\kappa$), and the Area Under the ROC (Receiver Operating Characteristic) Curve (AUC-ROC) for a comprehensive evaluation of classification effectiveness.

– **Accuracy** represents the ratio of correctly classified instances to the total dataset.

$$Accuracy = \frac{TP + TN}{TP + FP + FN + TN}$$

– **FPR** is the proportion of legitimate vehicles incorrectly classified as malicious.

$$FPR = \frac{FP}{FP + TN}$$

– **Recall** indicates the proportion of correctly predicted misbehaving instances of all received misbehaving messages.

$$Recall = \frac{TP}{TP + FN}$$

– **Precision** represents the proportion of correctly predicted attacks out of the dataset classified as an attack.

$$Precision = \frac{TP}{TP + FP}$$

– **F1-score** calculates the harmonic mean of precision and recall.

$$F1 - score = 2 \times \frac{Recall \times Precision}{Recall + Precision}$$

– **MCC** is a quality measure for binary classifications. Unlike Accuracy and F1-score, MCC considers the balance ratios of the 4 categories in the confusion matrix to accurately determine the proportion of correctly classified responses.

$$MCC = \frac{TP \times TN + FP \times FN}{\sqrt{(TP + FP)(TP + FN)(TN + FP)(TN + FN)}}$$

– **Cohen's Kappa** is a measure of positive agreement, which is similar to accuracy but agreement is subtracted with a probability of chance $P$.

$$P = \frac{(TP + FP) \times (TP + TN) + (TN + FP) \times (TN + FN)}{(TN + TP + FP + FN)^2}$$

$$\kappa = \frac{ACC - P}{1 - P}$$

– **AUC ROC**: plots true positive rate against false positive rate. Its values typically fall within the range of 0.5 to 1, with higher values indicating superior detection performance.

## 4.2   Performance Analysis

We evaluate the performance of misbehavior detection techniques for masquerade attacks in VANET networks using the J48, RF, MLP and SMO algorithms. The results of the classification presented in Table 4 indicate that all applied classifiers achieved high rates of accuracy, generally exceeding 90%, except for SMO, which had an accuracy of 87%. The RF classifier achieved the highest accuracy among the others, with 91%. For this same algorithm, we obtained the highest Cohen's Kappa (k) value close to 1, specifically equal to 0.6. This implies a strong agreement among categorical observers that is not due to chance. For the J48, MLP and SMO algorithms, we obtained a Kappa value of 0.5, which indicates moderate agreement between observers.

**Table 4.** Evaluation of masquerade attacks using ACC, $k$, and MCC metrics

| Metric | J48 | RF | MLP | SMO |
|---|---|---|---|---|
| *Correctly Classified Instances (ACC %)* | 90.294 | 90.993 | 90.224 | 87.009 |
| *Cohen's Kappa (k)* | 0.537 | 0.598 | 0.536 | 0.496 |
| *Matthews Correlation Coefficient (MCC)* | 0.785 | 0.81 | 0.795 | 0.729 |

We also use the Matthews Correlation Coefficient (MCC) to evaluate the detection of these 3 types of attack. Indeed, the F1 score is not particularly useful with highly unbalanced classes, especially when the class of interest is considered as a negative class but needs to be predicted with a high degree of precision. In our case, the "Genuine" class is the positive class, so a high F1 score does not necessarily indicate effective prediction of masquerade attacks. This is due to the asymmetrical nature of the F1 score, which does not give the same results when positive and negative classes are swapped. This is where the role of the MCC comes in, along with the computation of the F1 score for each class individually (see the F1 score values for each class and the weighted value in Table 5). The MCC values shown in Table 4 indicate that the RF algorithm effectively predicts masquerade attack classes compared to other models, with the closest value to 1, being 0.81. This implies a strong positive correlation with RF.

The Table 5 shows the evaluation results for each class independently [3], namely the "Genuine", "Dos Random Masquerade", "Data Replay Masquerade" and "DoS Disruptive" classes. For each class, various metrics were used, including FPR, recall, F1 score, precision, and AUC-Roc. The last rows of this table represent the weighted averages of these measurements.

Table 6, presented in the form of confusion matrices, provides a more detailed illustration of masquerade attack classification results in terms of true positives (TP), false negatives (FN), false positives (FP) and true negatives (TN) for the J48, RF, MLP and SMO algorithms applied to the Lust scenario dataset.

**Table 5.** Results of the evaluation with weighted metrics for masquerade attacks

|  | Evaluation Parameters | J48 | RF | MLP | SMO |
|---|---|---|---|---|---|
| Genuine | FPR | 0.145 | 0.171 | 0.125 | 0.011 |
|  | Recall | 0.985 | 0.996 | 0.987 | 0.938 |
|  | F1-score | 0.983 | 0.987 | 0.985 | 0.967 |
|  | Precision | 0.981 | 0.978 | 0.984 | 0.999 |
|  | AUC-Roc | 0.905 | 0.998 | 0.993 | 0.976 |
| Dos Random Masquerade | FPR | 0.012 | 0.002 | 0.029 | 0.032 |
|  | Recall | 0.572 | 0.723 | 0.674 | 0.263 |
|  | F1-score | 0.558 | 0.798 | 0.486 | 0.211 |
|  | Precision | 0.545 | 0.892 | 0.38 | 0.176 |
|  | AUC-Roc | 0.638 | 0.994 | 0.97 | 0.873 |
| Data Replay Masquerade | FPR | 0.023 | 0.025 | 0.022 | 0.049 |
|  | Recall | 0.258 | 0.15 | 0.139 | 0.408 |
|  | F1-score | 0.322 | 0.199 | 0.189 | 0.383 |
|  | Precision | 0.428 | 0.293 | 0.298 | 0.362 |
|  | AUC-Roc | 0.58 | 0.942 | 0.939 | 0.64 |
| DoS Disruptive Masquerade | FPR | 0.048 | 0.046 | 0.036 | 0.053 |
|  | Recall | 0.022 | 0.037 | 0.131 | 0.3 |
|  | F1-score | 0.016 | 0.027 | 0.107 | 0.184 |
|  | Precision | 0.012 | 0.021 | 0.09 | 0.133 |
|  | AUC-Roc | 0.229 | 0.89 | 0.897 | 0.756 |
|  | Weighted Avg.FPR | 0.131 | 0.154 | 0.114 | 0.015 |
|  | Weighted Avg.Recall | 0.903 | 0.91 | 0.902 | 0.87 |
|  | Weighted Avg.F1-score | 0.903 | 0.910 | 0.902 | 0.870 |
|  | Weighted Avg.Precision | 0.909 | 0.907 | 0.901 | 0.914 |

These results show that the RF algorithm outperforms all other algorithms for most of the evaluated parameters. This algorithm efficiently detects DoS Random Masquerade attacks, as well as genuine messages of the "Genuine" class. However, we observe a certain imbalance, or even inefficiency, of the models for detecting both attacks Data Replay Masquerade and DoS Disruptive Masquerade. This result aligns with our expectations and can be attributed to the challenge of distinguishing between normal and malicious messages in the Data Replay masquerade attack, as they both use the same pseudonym. Consequently, some malicious messages are classified as normal, and vice versa. Since the DoS Disruptive Masquerade type represents a combination of the first two types, it is also challenging to identify this type of attack. Overall detection of all these types of malicious behavior gives an accuracy of 91% and a false positive rate of 15.4% for the RF algorithm.

**Table 6.** Confusion Matrices

J48

| Genuine | DoSRandomMasquerade | DataReplayMasquerade | DoSDisruptiveMasquerade | Classified as |
|---|---|---|---|---|
| 34941 | 139 | 226 | 167 | Genuine |
| 146 | 580 | 44 | 244 | DoSRandomMasquerade |
| 239 | 196 | 655 | 1449 | DataReplayMasquerade |
| 285 | 150 | 606 | 23 | DoSDisruptiveMasquerade |

RF

| Genuine | DoSRandomMasquerade | DataReplayMasquerade | DoSDisruptiveMasquerade | Classified as |
|---|---|---|---|---|
| 35325 | 37 | 73 | 38 | Genuine |
| 186 | 733 | 35 | 60 | DoSRandomMasquerade |
| 415 | 30 | 382 | 1712 | DataReplayMasquerade |
| 188 | 22 | 815 | 39 | DoSDisruptiveMasquerade |

MLP

| Genuine | DoSRandomMasquerade | DataReplayMasquerade | DoSDisruptiveMasquerade | Classified as |
|---|---|---|---|---|
| 34997 | 105 | 211 | 160 | Genuine |
| 87 | 683 | 101 | 143 | DoSRandomMasquerade |
| 352 | 740 | 352 | 1095 | DataReplayMasquerade |
| 140 | 269 | 516 | 139 | DoSDisruptiveMasquerade |

SMO

| Genuine | DoSRandomMasquerade | DataReplayMasquerade | DoSDisruptiveMasquerade | Classified as |
|---|---|---|---|---|
| 33261 | 410 | 1155 | 647 | Genuine |
| 13 | 267 | 198 | 536 | DoSRandomMasquerade |
| 26 | 583 | 1035 | 895 | DataReplayMasquerade |
| 10 | 261 | 474 | 319 | DoSDisruptiveMasquerade |

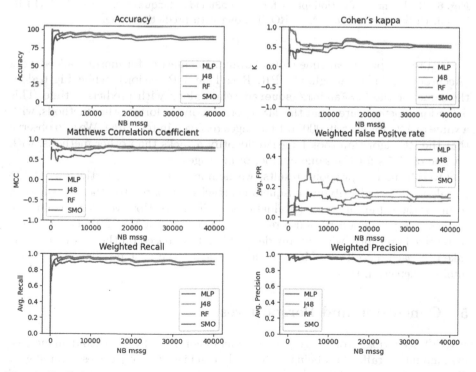

**Fig. 5.** Real-time evaluation metrics plots for masquerade attacks (Acc, $k$, MCC, weighted FPR, Recall, and Precision)

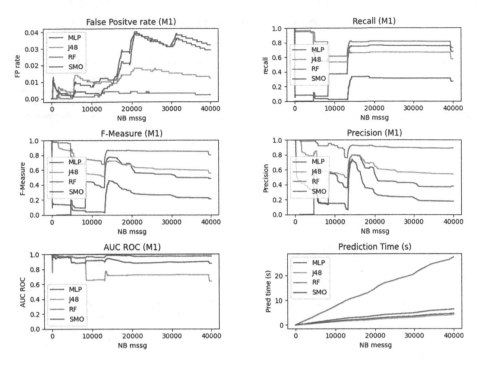

**Fig. 6.** Real-time evaluation plots for the `DoSRandomMasquerade` class (M1) (FPR, Recall, F1-Score, Precision, AUC ROC), along with prediction time.

Figure 5 displays real-time plots of evaluation metrics for masquerade attacks (Accuracy, $k$, MCC, weighted FPR, Recall, and Precision), while Fig. 6 does the same for the `DoSRandomMasquerade` class along with prediction time. This detection time indicates that the J48 algorithm outperforms other methods, with a value of 4.2 s for over 40,000 test messages received in real time. We also observe that the RF algorithm took longer to identify attacks than other methods, with a value of 27.5 s for the same number of messages.

Overall, in the previous results, we demonstrated the effectiveness of the RF algorithm in identifying masquerade attacks compared to other classifiers in terms of the values of various evaluation metrics used. However, for certain case studies, the prediction response related to security in a VANET network must be received promptly without any delay. Therefore, based on the prediction time taken by each algorithm, J48 detection would be the best possible choice with minimal detection time.

## 5    Conclusion and Perspectives

VANET scenarios involve a growing number of connected vehicles and an increasing amount of traffic data being exchanged, to achieve the objectives of Intelligent Transport Systems (ITS). This leads to a significant increase in the number and types of attacks or faulty behavior in the data transmitted. As a result, security measures need to be closely monitored to ensure that vehicles can communicate safely on the network. In this paper, we propose a study of a feasible attack that is prevalent in vehicular networks, known as Masquerade. For its implementation, we have identified three variants: Dos Random Masquerade, Data Replay Masquerade and DoS Disruptive Masquerade. We obtained high values for various measures using four algorithms (RF, J48, MLP, SMO), including accuracy and weighted measures such as F1 score, recall and precision. All these values are above 87%, which indicates that the models effectively distinguish between genuine and malicious messages. Regarding classification based on the type of attack, we found that RF effectively detects DoS Random Masquerade attacks as well as normal messages. However, the other two types of masquerade attacks are poorly detected, aligning with our expectations. Additionally, the RF algorithm delivers superior results in terms of accuracy, while J48 is quicker in classifying messages with minimal prediction time. To further improve our results, we plan to propose hybrid approaches that take advantage of various basic ML algorithms or implement neural network architectures.

# References

1. Arif, M., Wang, G., Bhuiyan, M.Z.A., Wang, T., Chen, J.: A survey on security attacks in VANETs: communication, applications and challenges. Veh. Commun. **19**, 100179 (2019)
2. Behrisch, M., Bieker, L., Erdmann, J., Krajzewicz, D.: SUMO–simulation of urban mobility: an overview. In: Proceedings of SIMUL 2011, The Third International Conference on Advances in System Simulation. ThinkMind (2011)
3. Chaouche, Y., Renault, É., Boussaha, R.: WEKA-based real-time attack detection for VANET simulations. In: 2023 International Conference on Software, Telecommunications and Computer Networks (SoftCOM), pp. 1–6. IEEE (2023)
4. Codeca, L., Frank, R., Engel, T.: Luxembourg SUMO Traffic (LuST) scenario: 24 hours of mobility for vehicular networking research. In: 2015 IEEE Vehicular Networking Conference (VNC), pp. 1–8. IEEE (2015)
5. Garner, S.R., et al.: WEKA: the waikato environment for knowledge analysis. In: Proceedings of the New Zealand Computer Science Research Students Conference, vol. 1995, pp. 57–64. Citeseer (1995)
6. Ghosal, A., Conti, M.: Security issues and challenges in V2X: a survey. Comput. Netw. **169**, 107093 (2020)
7. Hasbullah, H., Soomro, I.A., et al.: Denial of service (DOS) attack and its possible solutions in VANET. Int. J. Electron. Commun. Eng. **4**(5), 813–817 (2010)
8. Jo, H.J., Kim, J.H., Choi, H.Y., Choi, W., Lee, D.H., Lee, I.: MAuth-CAN: masquerade-attack-proof authentication for in-vehicle networks. IEEE Trans. Veh. Technol. **69**(2), 2204–2218 (2019)
9. Kamel, J., Ansari, M.R., Petit, J., Kaiser, A., Jemaa, I.B., Urien, P.: Simulation framework for misbehavior detection in vehicular networks. IEEE Trans. Veh. Technol. **69**(6), 6631–6643 (2020)
10. Khatri, S., et al.: Machine learning models and techniques for VANET based traffic management: implementation issues and challenges. Peer-to-Peer Netw. Appl. **14**, 1778–1805 (2021)

11. Rasheed, A., Gillani, S., Ajmal, S., Qayyum, A.: Vehicular ad hoc network (VANET): a survey, challenges, and applications. In: Laouiti, A., Qayyum, A., Mohamad Saad, M. (eds.) Vehicular Ad-Hoc Networks for Smart Cities, vol. 548, pp. 39–51. Springer, Singapore (2017). https://doi.org/10.1007/978-981-10-3503-6_4

12. Sharma, S., Kaul, A.: A survey on intrusion detection systems and honeypot based proactive security mechanisms in VANETs and VANET cloud. Veh. Commun. **12**, 138–164 (2018)

13. Sommer, C., et al.: Veins: the open source vehicular network simulation framework. In: Virdis, A., Kirsche, M. (eds.) Recent Advances in Network Simulation: The OMNeT++ Environment and Its Ecosystem. EAISICC, pp. 215–252. Springer, Cham (2019). https://doi.org/10.1007/978-3-030-12842-5_6

14. Varga, A., Hornig, R.: An overview of the OMNeT++ simulation environment. In: 1st International ICST Conference on Simulation Tools and Techniques for Communications, Networks and Systems (2010)

15. Zeng, Z.Q., Yu, H.B., Xu, H.R., Xie, Y.Q., Gao, J.: Fast training support vector machines using parallel sequential minimal optimization. In: 2008 3rd International Conference on Intelligent System and Knowledge Engineering, vol. 1, pp. 997–1001. IEEE (2008)

# Detecting Virtual Harassment in Social Media Using Machine Learning

Lina Feriel Benassou, Safa Bendaouia, Osman Salem[✉], and Ahmed Mehaoua

Centre Borelli, UMR 9010, Université Paris Cité, Paris, France
{linaferiel.benassou,safa.bendaouia,osman.salem,ahmed.mehaoua}@u-paris.fr

**Abstract.** The escalating prevalence of cyberbullying stands as a significant challenge in our digital era, causing severe repercussions on individuals' mental health, well-being, and professional integrity. This paper introduces an innovative approach to combat this online menace. We present a cyberbullying detection model based on machine learning principles. Our method involves curating a dataset comprising a variety of messages pre-labeled into multiple harassment categories. By leveraging sophisticated machine learning techniques, we extract distinctive features from these messages, enabling precise identification of cyberbullying instances. The results obtained underscore the high accuracy of our detection model, highlighting its undeniable effectiveness in combating online cyberbullying. This research thus offers a significant contribution by presenting an innovative method to detect and address this concerning phenomenon on digital platforms.

**Keywords:** Online harassment · Text Embedding · Machine learning · Social Media · Automated Learning

## 1 Introduction

Cyberbullying has deeply embedded itself within our digital era, causing devastating impacts on individuals' mental health and overall well-being. The evolution of online platforms and social networks has facilitated, expanded, and anonymized this scourge, leaving victims in a state of powerlessness, isolation, and an inability to counter recurrent attacks [18]. Consequently, the detection of cyberbullying has become a priority for researchers and online security professionals. The emergence of machine learning stands as a promising avenue to analyze extensive volumes of data and swiftly identify problematic messages.

In recent years, several approaches have been developed to detect cyberbullying online. Some traditional methods rely on predefined rules or word dictionaries to spot problematic messages [14]. However, the efficacy of these methods is constrained by the diverse forms of cyberbullying and the array of languages used. Advanced approaches, such as supervised and unsupervised classification [6,17], natural language processing (NLP) [15] techniques and anomaly detection [5], have surfaced.

É. Renault et al. (Eds.): MLN 2023, LNCS 14525, pp. 185–198, 2024.
https://doi.org/10.1007/978-3-031-59933-0_13

Supervised classification techniques leverage machine learning models to distinguish cyberbullying messages from normal ones, while unsupervised approaches use clustering or anomaly detection to identify problematic messages [4]. NLP techniques extract relevant features from online messages, such as tone, emotion, and context, encompassing sentiment detection, polarity analysis, and sarcasm analysis [9]. Recent studies have explored character sequence detection methods to pinpoint cyberbullying online, aiming to identify recurrent patterns in these messages.

Overall, online cyberbullying detection is continuously evolving, propelled by diverse approaches rooted in machine learning, natural language processing, and character sequence detection. Despite significant advancements, achieving accurate and reliable detection of online cyberbullying remains a major challenge, necessitating ongoing research and the development of novel methods. Our research aligns with this progression by introducing an innovative method to detect online cyberbullying.

The rest of this paper is structured as follows. In Sect. 2, we review recent work on for online detection of cyberbullying. In Sect. 3, we present the data collection, preprocessing, and feature extraction. In Sect. 4, we introduce our proposed approach for cyberbullying detection using the supervised Random Forest (RF) and Gradient Boosting (GB) algorithms. In Sect. 5, we present the experimental results of our approach and compare the detection accuracy of four supervised machine learning (ML) methods. Finally, in Sect. 6, we conclude the paper and discuss potential future research directions.

## 2   Related Works

The detection of online cyberbullying is a dynamic and evolving research field that has seen significant advancements in recent times. Various strategies have been developed to address cyberbullying, ranging from rule-based methods to machine learning-based approaches [12] and hybrid methods [1]. Rule-based methods, typically relying on predefined rules or dictionaries, have limitations, often failing to detect subtle forms of cyberbullying, such as sarcasm or irony, leading to a significant number of false positives.

On the other hand, machine learning methods employ algorithms to identify patterns and features within online messages indicative of cyberbullying. However, these methods necessitate a substantial volume of labeled data for training and can be susceptible to biases existing within the data [7]. Hybrid approaches, amalgamating rule-based techniques with machine learning methods, aim to harness the strengths of each approach. For instance, a hybrid method might utilize rule-based methods to detect explicit cases of cyberbullying while machine learning-based methods uncover more nuanced forms.

An increasing number of researchers have delved into online cyberbullying detection. Particularly, the work by Raj *et al.* in [16] introduced a cyberbullying detection method based on deep learning and sentiment analysis. Other research, such as that by Batani et al., utilizing deep learning models for cyberbullying detection on social media [2], and Mishra *et al.* in [11] with their real-time cyberbullying detection system, have also made significant contributions. Additionally, the study by Li *et al.* in [10] focused on cyberbullying detection based on sentiment polarity analysis.

Nevertheless, despite these remarkable advancements, the field of online cyberbullying detection faces several challenges. For instance, pinpointing cyberbullying within online conversations can be challenging due to the intricacy of these interactions. Furthermore, cyberbullying detection models might carry cultural and linguistic biases, resulting in detection errors or false positives.

In summary, online cyberbullying detection is a critical field requiring the resolution of technological, cultural, and legal challenges. Progress in this domain could substantially enhance online safety and prevent the proliferation of cyberbullying, a growing issue within our digital society.

## 3  Data Collection and Preprocessing

We choose to use a freely accessible dataset from the social networking site Twitter, which is a rich source of social data and can be easily downloaded from a public repository.

Our dataset consists of 39,869 tweets labeled according to the following categories: age, ethnicity, gender, religion, and not-bullying. The number of records in each category is presented in Table 1. These labels allow us to conduct a thorough analysis of different types of online harassment.

**Table 1.** Class distributions in dataset

| Type | Count |
| --- | --- |
| Age | 7992 |
| Ethnicity | 7961 |
| Gender | 7973 |
| Religion | 7998 |
| Not-bullying | 7945 |

It should be noted that collecting data from social networks can be complex due to the various characteristics and constraints associated with data collection. Therefore, we took care to select a representative sample of tweets that cover a wide range of topics and user profiles. We also took precautions to preserve the privacy of Twitter users by avoiding the collection of personally identifiable information such as usernames and email addresses.

In Fig. 1, we present a comprehensive visual representation of our detection system, which consists of a series of interconnected steps designed to efficiently process and analyze data for accurate detection and classification of cyberbullying.

The initial phase of our detection system involves data collection, where relevant information is gathered from various sources. This is a crucial foundation for the subsequent stages, ensuring that we have a comprehensive dataset to work with. Following data collection, we transition to the preprocessing and data cleaning stage. In this step, the collected data undergoes rigorous cleaning and preparation. Noise and inconsistencies are removed, and the data is transformed into a standardized format. This essential process sets the stage for accurate and reliable analysis in the subsequent steps.

The third step in our system focuses on feature extraction, a pivotal element in the success of our detection model. Here, we employ advanced techniques such as word embedding to convert the processed data into a structured format that highlights relevant features. These features are instrumental in capturing the underlying patterns and relationships within the data, setting the stage for effective classification.

With feature extraction completed, the system proceeds to the classification model. In this step, a machine learning model is trained on the extracted features, allowing the system to distinguish and categorize data into predefined classes or categories. The classification model is the core component responsible for making informed decisions based on the processed data.

The final step in our detection system is performance evaluation. This phase assesses the accuracy, reliability, and effectiveness of the classification model. It involves rigorous testing, validation, and benchmarking to ensure that the system meets the desired detection goals and objectives.

Text processing is essential for preparing a dataset for deep learning-based exploration and classification of online publications. To achieve this goal, the dataset must first be thoroughly purified and cleansed.

It is imperative to purify the datasets. To do so, remove HTML tags, convert all characters to lowercase, and remove non-alphabetic characters, URLs, mentions, and hashtags using the 'sub()' method from Python's standard library. Next, break the text into words using the 'word_tokenize()' method from the NLTK library and remove English language stop words using the 'stopwords.words('english')' method from the same library. Finally, stem each word using the 'PorterStemmer()' method from NLTK.

These preliminary phases are of paramount importance in enhancing the quality of the data by discarding unnecessary or redundant terms and reducing the word count. The result is a clean, well-structured dataset, poised for exploration in the context of a comprehensive analysis [3].

In various text mining projects, word embedding techniques are employed to create vector representations of the terms found within textual material. During text classification, software applications aim to derive real-valued word vectors. These vectors encapsulate the semantic meaning and contextual relationships of

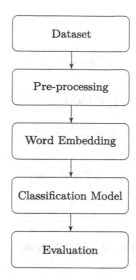

**Fig. 1.** Virtual harassment behavior detection system

nearby words in a vector space, enabling the anticipation that their meanings are alike. Based on this approach, our study focuses on the meticulous selection of vocabulary present in the dataset and its transformation utilizing the TF-IDF vectorization method [19].

TF-IDF, a well-established text processing technique, operates by converting a document into a numerical vector. This method relies on evaluating the frequency of each term's occurrence within the document. The key stages of this method encompass measuring the frequency of each specific term in the document, assessing the term's significance across the entire corpus, merging the Term Frequency (TF) scores with the Inverse Document Frequency (IDF) scores for each term, and finally, representing the document as a numerical vector where each term holds a specific TF-IDF score (as shown in Fig. 2).

The choice of this method proves astute owing to its ability to offer an efficient and concise representation of textual data. It takes into consideration both the frequency of a term in a specific document and its frequency across the entirety of the corpus. This approach aids in the identification of key terms and facilitates the differentiation among documents by focusing on their semantic and contextual content, thereby enhancing the understanding and analysis of the textual material.

## 4   Proposed Model

The next step is to categorize the social media postings into different sorts of online bullying categories after preprocessing the cyberbullying dataset. We test two different ensemble methods approaches for the classification tasks: RF and GB.

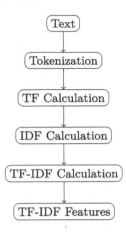

**Fig. 2.** TF-IDF system

We use the RF algorithm to train our prediction model. The RF model is a supervised learning algorithm that uses a set of decision trees to classify or predict data. The model builds a large number of decision trees (shown in Fig. 3) on random samples of training data. When new data is presented to the model, it is passed through each tree and a prediction is made. Then, the predictions of each tree are combined to give a final prediction for the data. RF is known for its high accuracy, ability to handle complex data, and detect interactions between attributes, which can lead to better predictions. Additionally, the model can handle missing or noisy data without requiring additional data cleaning [13].

We use the default parameters of the RF algorithm, as provided by the scikit-learn machine learning library. These default parameters are presented in Table 2. The default RF algorithm uses 100 decision trees for classification or prediction, uses the Gini index to measure the quality of the tree split, and does not impose a maximum limit on the depth of the tree. The minimum number of samples required to split an internal node is 2, while the minimum number of samples required to be a leaf is 1. Additionally, the algorithm considers all features to find the best tree split, uses samples with replacement (bootstrap = True), and uses a random seed for reproducibility of results.

We also use the GB algorithm to train our prediction model. GB is a supervised learning algorithm that builds an ensemble of weak prediction models, typically decision trees, in a sequential manner as shown in Fig. 4. Each model tries to correct the errors made by the previous one, leading to a final strong predictor. The algorithm works by minimizing a loss function that measures the difference between the predicted values and the true values of the training data.

We use the scikit-learn library's GB classifier with n_estimators = 100 and learning_rate = 0.1, and the used parameters in our model are presented in Table 3. The n_estimators parameter determines the number of decision trees to be used in the model. In our case, we used 100 trees. The learning_rate parameter controls the contribution of each tree to the final prediction. A smaller learning

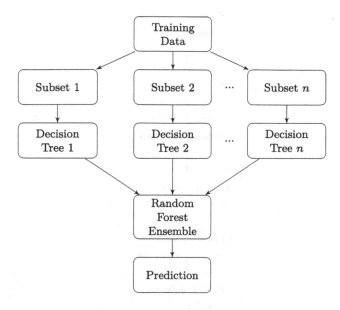

**Fig. 3.** Random Forest Algorithm

**Table 2.** Summary of the parameters used in the RF model

| Parameter Name | Value |
| --- | --- |
| n_estimators | 100 |
| criterion | gini |
| min_samples_split | 2 |
| min_samples_leaf | 1 |
| min_weight_fraction_leaf | 0.0 |
| max_features | auto |
| min_impurity_decrease | 0.0 |
| oob_score | False |
| warm_start | False |
| ccp_alpha | 0.0 |

rate means each tree has a smaller influence, while a larger learning rate means each tree has a larger influence.

The GB is known for its high accuracy and its ability to handle complex data with interactions between attributes. It can also handle missing or noisy data without requiring additional data cleaning. However, it can be computationally expensive and sensitive to overfitting if the number of trees is too high or the learning rate is too low.

To evaluate the performance of our classification model, we use the standard metrics recommended in the literature, including precision, recall, F1-score, and

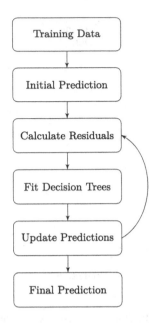

**Fig. 4.** Gradient Boosting Algorithm

accuracy. Precision measures the proportion of true positives among positive predictions, which evaluates the ability of our model to provide reliable results. Recall measures the proportion of true positives detected among all real positive cases, which is important to evaluate the ability of our model to detect all positive cases, even if they are rare. F1-score is a measure of the combined precision and recall in a single value, which allows for the comparison of the performance of different classification models. Finally, accuracy measures the proportion of correct predictions among all predictions, which is an important criterion to evaluate the ability of our model to correctly classify the data. These standard metrics are widely used in the scientific community for the evaluation of classification models, which allows us to compare our results with those of previous work in the field.

## 5   Experimental Results

### 5.1   Results

This section presents the findings of our studies on a multiclass classification using ensemble methods developed for a cyberbullying detection system. It includes information about bullying, age, religion, ethnicity, gender, and not-bullying. To partition the dataset samples into training, and testing sets before executing the classification job, as shown in Table 4.

**Table 3.** Summary of the parameters used in the GB model

| Parameter Name | Value |
|---|---|
| n_estimators | 100 |
| learning_rate | 0.1 |
| loss | deviance |
| subsample | 1.0 |
| criterion | friedman_mse |
| min_samples_split | 2 |
| min_samples_leaf | 1 |
| min_weight_fraction_leaf | 0.0 |
| max_depth | 3 |
| min_impurity_decrease | 0.0 |
| alpha | 0.9 |
| warm_start | False |
| presort | deprecated |
| validation_fraction | 0.1 |
| tol | 0.0001 |

**Table 4.** Splits of dataset

| Dataset Name | Total of Samples | Training Set 80% | Testing Set 20% |
|---|---|---|---|
| Cyberbullying tweets | 39869 | 31,895 | 7974 |

In the evaluated multiclass dataset, our models were examined to identify and categorize several cyberbullying classes (gender, ethnicity, age, religion, and non-bullying). We use a variety of supervised classification and assessment measures, such as precision, recall, F1-score, and accuracy, to evaluate the models. Figures 5 and 6 illustrate the confusion matrices used to calculate these measurements.

Out of the 7974 samples used in the testing set, Fig. 5 above shows that 468 tweets are classified incorrectly samples using the RF model for all classes. We observe that the model encounters some difficulties in categorizing the gender class.

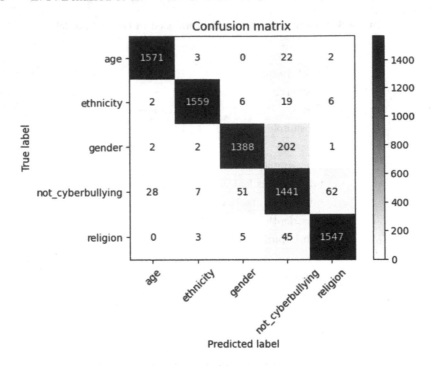

**Fig. 5.** Confusion matrix of the RF model

The GB misclassified 514 tweet samples when comparing the performance of the models about the misclassification rate, as shown in Fig. 6. We observe that the model encounters the same difficulties as RF with categorizing the gender class. It may be due to the features chosen to represent this class, it may not be discriminative enough to allow the model to correctly distinguish gender from not-bullying. As a result, the RF model and GB are the same in terms of classification accuracy. Table 5 shows the experimental findings for the multiclass classification dataset: Table 5 illustrates the Results of the RF and GB models. For the GB model, we found that the classes "age", "ethnicity", and "religion" have a precision of 99%, 100%, and 97%, recalls of 97%, 98%, and 94%, and F1-score of 98%, 99%, and 96%, respectively. This indicates that our model did a great job of predicting the age, ethnicity, and religious types of cyberbullying that were present in our dataset. Our model achieves an F1-score of 90%, a precision of 96%, and a recall of 85% for the "gender" class.

Although recall is low, precision is good, indicating that the model might have trouble correctly classifying some gender types. Our model scored 79% precision, 93% recall, and an F1-score of 86% for the "not-Bullying" class. recall is high despite the low precision, indicating that our model can effectively identify comments that are not bullying. We observe that we have almost the same results with RF.

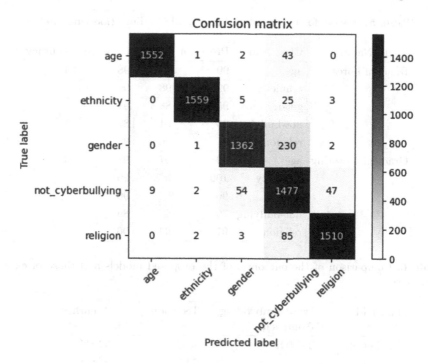

**Fig. 6.** Confusion matrix of the GB model

## 5.2    Comparison

The results of the ensemble methods utilizing the multiclass cyberbullying classification dataset are compared with those of the Wang *et al.* in [21], Talpur *et al.* in [20] and Hani *et al.* in [8]. Table provides an overview of the findings using the same dataset with accuracy as a metric.

As shown in Table 6, the TF-IDF method for word representation was utilized in the initial publication by Hani *et al.*, in addition to deep learning methods including Artificial Neural Networks (ANNs) and Support Vector Machines (SVMs). For ANN and SVM, respectively, they attained an accuracy of 92.8% and 90.3%.

In the second paper, Talpur *et al.* used the RF technique in conjunction with the Word2vec approach for word representation. They had a 93% accuracy score.

The third study was written by Wang *et al.* and used (CGN) and XGBoost with Word2vec for word representation. For GCN, they achieved 87% accuracy, and for XGBoost, 94%.

Our research leverages the robustness of TF-IDF for word representation, coupled with RF and GB techniques. This combination results in a remarkable 94% accuracy rate. The synergy of TF-IDF's term significance capture and the predictive strength of ensemble methods makes our approach stand out.

In conclusion, our research highlights the effectiveness of employing the TF-IDF method in conjunction with machine learning techniques like RF and GB.

**Table 5.** Results for the Random Forest and Gradient Boosting models

| Model Name | Class Name | Precision | Recall | F1-score | Accuracy |
|---|---|---|---|---|---|
| Random Forest | age | 99 | 98 | 98 | 94 |
| | ethnicity | 99 | 98 | 99 | |
| | gender | 95 | 88 | 91 | |
| | not-bullying | 83 | 91 | 87 | |
| | religion | 95 | 96 | 96 | |
| Gradient Boosting | age | 99 | 97 | 98 | 94 |
| | ethnicity | 100 | 98 | 99 | |
| | gender | 96 | 85 | 90 | |
| | not-bullying | 79 | 93 | 86 | |
| | religion | 97 | 94 | 96 | |

**Table 6.** Comparison of the outcomes of the proposed models and those of existing techniques

| Paper Id | Word Embedding Approach | Technique | Accuracy |
|---|---|---|---|
| Hani et al. | TF-IDF | ANN | 92.8% |
| | | SVM | 90.3% |
| Talpur et al. | Word2vec | RF | 93% |
| Wang et al. | Word2vec | GCN | 87% |
| | | XGBoost | 94% |
| Proposed work | TF-IDF | RF | 94% |
| | | GBoost | 94% |

What distinguishes our approach is the careful selection of methods that complement each other. TF-IDF excels in representing the importance of terms, while RF and GB capitalize on this rich representation to make accurate classifications. This synergy and meticulous combination of methods have consistently yielded promising results in classifying online harassment comments.

## 6   Conclusion

Cyberbullying involves the use of digital platforms to intimidate, threaten, or pressure others, often taking place across diverse online spaces like social media, forums, or email. It's the unwelcome guest of online behavior, known for its use of harsh, aggressive, or menacing language. This study was geared towards crafting an effective cyberbullying detection tool, aimed at identifying and probing instances of online abuse occurring within the realms of social media. To refine and assess the proposed method, we rigorously trained and tested our system using versatile datasets employing multiclass classification.

Our pursuit encompassed a quest to categorize social media content, seeking patterns in bullying behaviors tied to gender, religion, ethnicity, age, cyberbullying, and not-bullying communication. We meticulously compared the accuracy of RF and GB classifiers, achieving an impressive 94% detection rate. Nonetheless, as with any journey, our study wasn't devoid of challenges.

One notable challenge we encountered was related to the "not-bullying" class. Despite the overall high accuracy, this particular class posed difficulties. The reason behind this challenge was the quality of the training examples. They weren't fully representative and failed to encompass the diversity of not-bullying communications.

Furthermore, it's essential to acknowledge that our model cannot eliminate the risks of false positives and false negatives. Subtleties in language, such as sarcasm, metaphors, and expressions, pose challenges to accurate detection. Moreover, the evolution of online behaviors, where users adapt their communication styles and modes in response to trends, can give rise to new forms of cyberbullying.

This experience underscores the importance of using comprehensive and representative datasets to foster a deeper understanding of online behaviors. Looking ahead, our future endeavors will be directed towards the development of a model capable of identifying cyberbullying within multilingual datasets, a crucial step towards a more inclusive and effective cyberbullying detection system.

# References

1. Abdullah, B., Murshed, H., Abawajy, J., Mallappa, S., Saif, M.A.N., Al-Ariki, H.D.E.: DEA-RNN: a hybrid deep learning approach for cyberbullying detection in twitter social media platform. IEEE Access **10**, 25857–25871 (2022)
2. Batani, J., et al.: A review of deep learning models for detecting cyberbullying on social media networks. In: Silhavy, R. (ed.) Cybernetics Perspectives in Systems. Lecture Notes in Networks and Systems, vol. 503, pp. 528–550. Springer, Cham (2022). https://doi.org/10.1007/978-3-031-09073-8_46
3. Bazeley, P.: Qualitative Data Analysis: Practical Strategies (2013)
4. Berry, M.W., Mohamed, A., Yap, B.W. (eds.): Supervised and Unsupervised Learning for Data Science (2019)
5. Cheng, L., Shu, K., Wu, S., Silva, Y.N., Hall, D.L., Liu, H.: Unsupervised cyberbullying detection via time-informed gaussian mixture model (2022). submitted on 6 Aug 2020
6. El-Seoud, S.A., Farag, N., McKee, G.: A review on non-supervised approaches for cyberbullying detection. IJEP **10**(4), 25–34 (2020)
7. Gomez, C.E., Sztainberg, M.O., Trana, R.E.: Curating cyberbullying datasets: a human-AI collaborative approach. Int. J. Bullying Preven. **4**, 35–46 (2021)
8. Hani, J., Nashaat, M., Ahmed, M., Emad, Z., Amer, E., Mohammed, A.: Social media cyberbullying detection using machine learning. Int. J. Adv. Comput. Sci. Appl. **10**(5), 1–8 (2019)
9. Kulkarni, A., Shivananda, A.: Natural Language Processing Recipes (2019). https://doi.org/10.1007/978-1-4842-4267-4

10. Li, X., Yang, F., Li, J., Li, Y., Wang, S.: Cyberbullying detection based on principal component analysis and logistic regression. In: Proceedings of the 3rd International Conference on Education, Culture and Social Development, pp. 183–187 (2019)

11. Mishra, R., Kuriakose, J., Joshi, A.: Bullyalert: a real-time cyberbullying detection system for Twitter. In: Proceedings of the Tenth ACM Conference on Web Science (2019)

12. Neelakandan, S., Sridevi, M., Chandrasekaran, S., Singh Pundir, A.K., Lingaiah, T.B.: Deep learning approaches for cyberbullying detection and classification on social media (2022). academic Editor: Akshi Kumar

13. Pedregosa, F., et al.: Scikit-learn: machine learning in python. J. Mach. Learn. Res. **12**, 2825–2830 (2011)

14. Perera, A., Fernando, P.: Accurate cyberbullying detection and prevention on social media. Procedia Comput. Sci. **181**, 605–611 (2021). https://doi.org/10.1016/j. procs.2021.01.207. under a Creative Commons license

15. Raj, C., Agarwal, A., Bharathy, G., Narayan, B., Prasad, M.: Cyberbullying detection: hybrid models based on machine learning and NLP techniques. Electronics **10**(22), 2810 (2021). https://doi.org/10.3390/electronics10222810

16. Raj, M., Singh, S., Solanki, K., Selvanambi, R.: An application to detect cyberbullying using machine learning and deep learning techniques. SN Comput. Sci. **3** (2022). article number: 401

17. Raza, M.O., Memon, M., Bhatti, S., Bux, R.: Detecting cyberbullying in social commentary using supervised machine learning. In: Advances in Intelligent Systems and Computing. Advances in Intelligent Systems and Computing, vol. 1130 (2020). first Online: 13 February 2020

18. Rodriguez-Rivas, M.E., Varela, J.J., González, C., Chuecas, M.J.: The role of family support and conflict in cyberbullying and subjective well-being among chilean adolescents during the COVID-19 period (2022). received 20 September 2021; Received in revised form 26 November 2021; Accepted 30 March 2022; Published by Elsevier Ltd. This is an open access article under the CC BY-NC-ND license

19. Salton, G., McGill, M.J.: Introduction to Modern Information Retrieval. McGraw-Hill, New York (1983)

20. Talpur, B.A., O'Sullivan, D.: Multi-class imbalance in text classification: a feature engineering approach to detect cyberbullying in Twitter. Informatics **7**(4) (2020). https://www.mdpi.com/2227-9709/7/4/52

21. Wang, J., Fu, K., Lu, C.: SOSNET: a graph convolutional network approach to fine-grained cyberbullying detection. In: Proceedings of the 2020 IEEE International Conference on Big Data (Big Data), pp. 1699–1708 (2020)

# Leverage Data Security Policies Complexity for Users: An End-to-End Storage Service Management in the Cloud Based on ABAC Attributes

Nicolas Greneche[1]([⊠]), Frederic Andres[2], Shihori Tanabe[3], Andreas Pester[4], Hesham H. Ali[5], Amgad A. Mahmoud[6], and Dominique Bascle[7]

[1] University of Sorbonne Paris North, LIPN - UMR 7030, 99 Avenue JB Clément, 93430 Villetaneuse, France
nicolas.greneche@univ-paris13.fr
[2] Digital Content and Media Sciences Research Division, National Institute of Informatics, 2-1-2 Hitotsubashi, Chiyoda-ku, Tokyo 101-8430, Japan
andres@nii.ac.jp
[3] Division of Risk Assessment, Center for Biological Safety and Research, National Institute of Health Sciences, 3-25-26 Tonomachi, Kawasaki-ku, Kawasaki 210-9501, Japan
stanabe@nihs.go.jp
[4] Faculty of Informatics and Computer Science, AI Group, The British University in Egypt, Suez Desert Road, Cairo, P.O. Box 43, El Sherouk City 11837, Egypt
andreas.pester@bue.edu.eg
[5] College of Information Science and Technology, University of Nebraska at Omaha, Omaha, NE 68182-0116, USA
hali@unomaha.edu
[6] Faculty of Informatics and Computer Science, Artificial Intelligence Department, The British University in Egypt, Suez Desert Road, El Sherouk City 11837, Cairo, Egypt
amgad.abdallah@bue.edu.eg
[7] University of Sorbonne Paris North, DSI - UNIF, 99 Avenue JB Clément, 93430 Villetaneuse, France
dominique.bascle@univ-paris13.fr

**Abstract.** This position paper presents a method to ease the management of data security from the user point of view. Nowadays, users have many ways to access the same data: direct connection to the host, shared filesystem or web drive-like solutions. This leads to complex data access control policies. At the same time, users have more and more liberty in resource instantiation. They can benefit from various self service storage facilities from many Cloud operators in an on-premise or remote way. Moreover, interfaces with these providers are designed in a way that real locations of data are hidden to give an illusion of infinite resources availability. Obviously, Cloud providers have many ways to fine tune resource allocation but users may not be aware of it. With this growth of resource distribution, access control also evolved. Formerly, a simple access control

É. Renault et al. (Eds.): MLN 2023, LNCS 14525, pp. 199–217, 2024.
https://doi.org/10.1007/978-3-031-59933-0_14

scheme based on identity was sufficient for data security (IBAC). With the complexity increase of access control, new schemes emerged based on roles (RBAC) or attributes (ABAC). We will investigate the last one because attributes rules access control but it also gives information on a user's profile that may be used to ease the creation and configuration of data services on distributed resources such as Cloud providers.

**Keywords:** Storage · Data · Security Policies · Access Control · ABAC

# 1    Introduction

With the advent of the Cloud model, users can dispose of elastic resources. They can benefit from horizontal or vertical scaling of their resources. However, this elasticity has a cost in terms of control loss on underlying infrastructure when it comes to public, hybrid or multi Cloud contexts. Users can regain control but it requires heavy skills on Cloud orchestrator configuration as well as a good knowledge of labeling policy of each Cloud provider. Without such knowledge, users have to rely on provider choices that may not meet technical or organizational requirements. Moreover, each Cloud provider has its own specificities regarding configurations and capabilities.

At the same time, Cloud adoption requires improvements on access control schemes. Regular access control based on identity does not scale well in dynamic environments as a new subject or a new object requires an update or access control rules to take the in account. As a consequence, a new access control paradigm emerged based on attributes. In this access control, users and objects have attributes and access control rules rely on them. That makes access control able to scale with resources. It also provides a fine grained access control on data themselves.

Our approach aims at relying on attributes expressivity to empower users with their storage oriented services. Attributes draw a profile of users that can be translated to security policies for their resources instantiation. As attributes were designed to provide fine grained access control in dynamic environments. The combination of native usage of attributes and attached user profile leads to an end-to-end security policy for data storage services.

This paper will give a review of access control schemes as well as security management issues on the Cloud environment. We will discuss Kubernetes as it is the most widespread orchestrator in Cloud environments. Then we will describe our proposal and illustrate it with two use cases. The first use case is quite generic and concerns a web drive service. This generic use case is divided into two parts. The first part illustrates the benefits of our approach, even with a non attributes aware application. The second part presents a full attribute aware deployment providing end-to-end security of the application. The second use case presents a specific application to Geom-SAC project which is a live project implemented for drug discovery, and molecular generation based on chemical, and geometrical attributes by utilizing an extension of Deep Reinforcement Learning and Graph Neural Networks algorithms, more on the environment of this approach for better understanding how it operates based on molecular attributes we direct the reader to [28].

## 2 Related Works

This section will be divided into two subsections. In the first subsection, we will explore the main access control schemes. This subsection will give the major outlines of each access control family. We will evaluate the benefits and caveats of each one. We will also show how they can be combined to handle the growth of security policies complexity. An access control policy is a permanent trade-off between maintainability, applicability and realistic usage of the system. That's why access control schemes can be stacked to achieve this trade-off.

The second subsection will discuss some scheduling strategy in Kubernetes. The first layer of every security policy is the protection offered by the place where your data belong. Formerly, it was closely related to physical security of the local server room. In the current Cloud environment, tables changed. Components of the same infrastructure can be spread on different datacenters in different countries with different levels of security and safety (tier 1 to 4). Physical security matters progressively disappeared in favor of container scheduling issues, specifically placement constraints regarding security objectives. Another concern regarding resource scheduling policy is the shifting from virtual or real operating system hardening to container engine security.

### 2.1 Access Control

In this subsection dedicated to access control we will first discuss how objects and subjects are handled following Discretionary Access Control (DAC) and Mandatory Access Control (MAC) paradigms. Then, we will describe three access control schemes: Identity Based Access Control (IBAC), Role Based Access Control (RBAC) and Attributes Based Access Control (ABAC). Access control rules what the subjects can do with objects. Subjects are mainly users and objects can be both low level entities such as files, network sockets etc. and high level entities like containers, nodes etc. DAC and MAC defines how subjects and objects are handled. In [15] authors make a review of DAC, MAC, RBAC and OrBAC.

### 2.1.1 DAC

Discretionary Access Control considers that each object has an owner that sets up an Access Control List (ACL) on it. An ACL is composed of Access Control

Entry (ACE). Each ACE is an access rule of a subject on an object. The "Discretionary" in DAC means that ACEs on objects are at the owner's discretion. There are several models of DAC. We will discuss Lampson, HRU and TAM.

*The Lampson Model* [25]. This model defines an access control matrix. Each line is a subject, eac column is an object and each cell is filled with a set access rights (mainly read, write and execute). In the Lampson model, updating policy is a bit tedious because each new subject (user) comes with an update of the access control matrix.

*The HRU (Harrison Ruzzo Ullman) model* [21] is an evolution of Lampson model. In this model, each subject is also an object. As a consequence, it enables subjects to modify the access control matrix. The leads to a splitted access control with an administrative set of rules (subjects on subjects) and access rules (subjects on objects). In [21] authors proved that in mono-operational protection system (i.e where all command contain only one atomic operation), safety problem is décidable. However, verification is NP-complete. Consequently, this model is unusable in operating systems. For example, creation of a file in a UNIX based operating system requires two operations: 1) object creation, then 2) access rights assignment.

*The TAM (Typed Access Matrix) model* [34] is an evolution of HRU. Authors add the notion of strong types that extends former works on SPM (Sandhu's Schematic Protection Model) [35]. Strong types are immutable labels on subjects and objects. TAM adds a finite set of types to HRU (addition and deletion of types are not allowed) as well as some management rules. In TAM, when objects are created, its type is fixed and never modified. Authors show that this model is decidable in its monotonic flavor (i.e. that do not allow creation or suppresison). This led to a new model called MTAM (Monotonic Typed Access Matrix) that removed TAM's suppression operations. However, the safety problem remains NP-complete. That's why authors extended NTAM to ternary MTAM where every command has three arguments at most. Consequently, safety in ternary MTAM has a complexity that can be solved in a polynomial time.

In this discussion, we saw that DAC's main asset is that it is fully decentralized. However, the security policy evolves in an unpredictable way for site administrators [14]. Operating Systems that only rely on DAC models are prone to privilege escalation (attackers that override DAC to gain administrative privileges). In [7,18,27], authors pinpoint the weaknesses of DAC models. The main issue is that users can define security policies on their owned objects. They can apply rules that may endanger confidentiality and integrity of the operating system. The MAC (Mandatory Access Control) is a complementary model aiming to confine the omnipotent administrator user. As stated in reports [7,22] a system that combines DAC and MAC models is far more secure than DAC-only.

## 2.1.2  MAC

MAC stands for Mandatory Access Control. Mandatory means that the access control is enforced for every user including the administrator. A MAC policy is an access control matrix enforced by a reference monitor. The security policy is fixed at system startup and cannot be modified during its execution. Thanks to this immutable access control matrix, it is possible to guarantee some access rules between subjects and objects. We will discuss three MAC models: Bell-LaPudala (BLP), Biba and Domain Type Enforcement (DTE).

*Bell-LaPudala (BLP)* [3] is a confidentiality oriented model designed for military purposes. It relies on the HRU model and excludes all object creation or suppression. Every subject and object of the operating system is labeled with a confidentiality level. Those labels enable two security properties:

1. **No Read Up** When a subject requires read access on an object, subject's confidentiality level must be superior or equal to object's confidentiality level;
2. **No Write Down** When a subject requires write access on an object, the subject's confidentiality level must be inferior or equal to the object's confidentiality level.

Some operating systems implemented the BLP model: MULTICS, Solaris or HP UX. In these systems, BLP is often referred to as MLS for Multi-Level Security. However, this model is difficult to implement in real life operating systems. The model needs some adaptation that leads to raise confidentiality levels of objects.

*Biba* [4] is always qualified of *dual* regarding BLP because it deals with integrity. Every subject and object of the operating system is labeled with an integrity level. Those labels enable two security properties:

1. **No Read Down** A subject is not allowed to read an object with a strictly inferior integrity level, this could corrupt subject's integrity;
2. **No Write Up** A subject can not write an object with a strictly superior integrity level.

The integration of Biba is almost impossible because each object would be on the bottom of the integrity scale.

*Domain and Type Enforcement (DTE)* [5] is a high level access control model available in some regular operating systems [2]. DTE is close to strong types defined in TAM. DTE enables implementation of other security models such as BLP or Biba. DTE aims at:

- Restraining available resources, especially for programs (subjects) running with administrative privileges. This is the least privilege principle defined in [29];
- Apply a strict control on programs (subjects) that require access to sensitive data. It emphasizes isolation.

DTE model reuses matrix access control as previous models, except that subject and object are replaced by domains and types. It means that each process runs in a given domain and each object (file, socket, shared memory etc.) has a type. DTE defines three tables:

- **Type table** that map types to objects;
- **Domain Definition Table (DDT)** that specifies domain's access rights on types;
- **Domain Interaction Table (DIT)** that specifies access rights between domains.

Those three tables are immutable during the execution of the operating system. A popular DTE implementation is SELinux. In [10], authors give an example of a SELinux policy for Apache web server illustrated in Fig. 1. In this policy, web developer is mapped to webdev_d domain, he has read and write access to Apache configuration. Apache service program mapped to apache_d has read access to its configuration (apache_conf_t) and web pages directory (apache_html_t). It is also able to read and write its temporary directory (apache_tmp_t) and execute shared libraries (apache_lib_t). Finally, web administrator mapped to webadmin_d can edit web pages (apache_html_t).

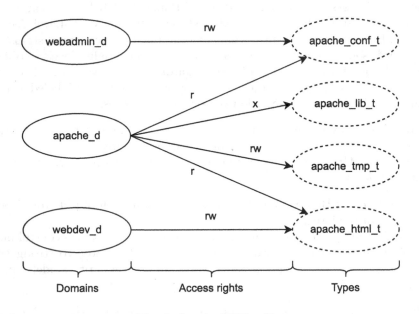

**Fig. 1.** Apache DTE policy

In this section, we enlight that MAC addresses DAC issues. There are two approaches for MAC policies. BLP or Biba models try to guarantee specific security properties while DTE provides a set of immutable access rules between subjects and objects through a labeling policy. However, in [8] authors present a method to guarantee generic security properties on top of DTE access control. They applied their method to several domains: Honeypots [9], HPC [19] and Cloud [6].

### 2.1.3    ABAC

In the previous section, we discussed access control models based on user identity. This category of access control is called IBAC that stands for Identity Based Access Control. IBAC categorizes almost all previous access control models. The main issue with this model is that it does not scale well. Another access control model based on roles emerged. This model is called Role Based Access Control (RBAC) [16,17]. In this model, users can embrace a role. As shown in Fig. 2, RBAC relies on the role to grant access to an object. NIST standardized RBAC in [32]. The role indirection between users and objects partially leverage security policies complexity for medium scale organizations. However, as stated in [26,31], RBAC needs some improvements to scale, especially in multi-organizations context. In [38], authors introduce ROBAC (Role and Organization Based Access Control) and in [12], authors present OrBAC (Organization Based Access Control). Both models include organization context to role to leverage scalability issues of RBAC. RBAC does not aim to replace existing models. Indeed, RBAC can model both DAC [33] and MAC [30].

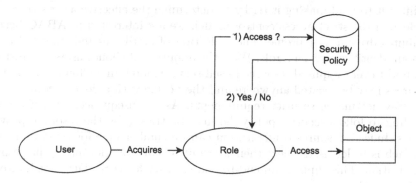

**Fig. 2.** RBAC

Another approach for access control scalability is proposed thanks to an Attribute Based Access Control (ABAC) [11]. ABAC uses attributes to perform access control between subjects and objects. The Fig. 3 presents the simplified architecture of ABAC. Both users and objects have attributes. These attributes are fixed by administrators, this is why ABAC is referenced as a centralized access control model. When a subject user tries to access an object, the request is intercepted by the Policy Enforcement Point (PEP). PEP is the mandatory access point to the object. The PEP forwards access attempt to the Policy Decision Point (PDP). The PDP retrieves subject and object attributes and confronts them to the security policy. Then, it informs PEP of the acceptance or rejection of the access.

An ABAC access rule has a subject attribute, an object attribute, an interaction and a context. Consequently, there is no need to add rules when new

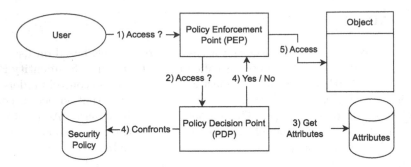

**Fig. 3.** ABAC

subjects or objects are created. They only require some attributes. This model makes it possible to have fine-grained access rules. In [23], authors present an implementation of ABAC in MySQL. In [20] authors deal with NoSQL database MongoDB. The challenge with ABAC is attribute maintenance. Administrators must be strict with attributes nomenclature. Moreover, the security policy may be difficult to read making it tricky to determine the effective access rules.

Beyond the strict access control model, we are interested in ABAC because attributes draw a user profile. The collection of attributes tells more of a user than an identifier or even a role. With the adoption of Cloud, users can instantiate almost every required resource in self-service mode. In a Cloud environment, resources may be located anywhere and they come with a default configuration that may not match security requirements. As a consequence, even if the user is willing to follow security policy, he may be tricked by the resource provider choices. Indeed, resource configuration can be cumbersome for most users. Our approach is to let users select their attributes that are relevant to the resource instantiation. This supplies an end-to-end security for their data, not limited to access control.

## 2.2 Containers Instantiation Security Complexity in Kubernetes

IT security has always been a concern of least and separation of privileges. In [13], written in 1976, author gives the guidelines to ensure security in operating systems. The keypoints of this paper are:

- *Least privilege*, i.e. an entity must have minimal access rights to perform its actions, not more;
- *Privileges separation*, i.e. an entity should only have one function (to ease the application of least privilege).

Prior to Cloud adoption, it concerned application's implementations, operating systems hardening and, of course, physical security of servers rooms. Organizations had control on each mentioned layer. As stated in [36], in Cloud environment security concerns moved from full stack security management to shared off-the-shelf remote components. In this section, we will give a short introduction to Kubernetes. Then, we will discuss complexity of security configuration of containers instantiation process.

### 2.2.1 Kubernetes

Kubernetes, commonly stylized as K8s [24] is an open-source container orchestration system for automating software deployment, scaling, and management. Kubernetes aimed to solve an entirely different problem than the traditional problems solved by HPC clusters – delivering scalable, always-on, reliable web services in Google's Cloud. Kubernetes applications are assumed to be containerized and adhere to a cloud-native design approach. Pods which are groups of one or more CRI-O[1] or OCI[2] compliant containers, are the primary constituents of applications that are deployed on a cluster to provide specific functionality for an application. Kubernetes provides features supporting continuous integration/delivery (CI/CD) pipelines and modern DevOps techniques. Health checks give mechanisms to send readiness and liveness probes to ensure continued service availability. Another differentiating feature is that Kubernetes is more than just a resource manager; it is a complete management and runtime environment. Kubernetes includes services that applications rely on, including DNS management, virtual networking, persistent volumes, secret keys management, etc.

### 2.2.2 Security Configuration of Containers Instantiation Process

Cloud providers Googe[3], AWS[4] and Azure[5] give their guidelines regarding security. These three providers rely on Kubernetes for container orchestration. In [1], authors pinpoint that Kubernetes implements least and separation of privileges by design. Kubernetes architecture segregates each of its functionality in a container. Each container only has access rights regarding its needs. Security issues come mainly from user's misconfigurations.

Indeed, container instantiation can take a large amount of parameters. For instance, users can decide on network exposition of their container with tailored filtering policies, they can choose an hardened container engine [37] such as gVisor or Katacontainers, they can use a namespace for isolation etc. In addition to these container related parameters, they may have to deal with higher level requirements regarding data security such as: specifications of hosting datacenter (especially service level and/or geographic location) or neighborhood concerns in scheduling (not sharing the node with un allowed containers). These constraints

---

[1] https://cri-o.io/.

[2] https://opencontainers.org/.

[3] https://cloud.google.com/kubernetes-engine/docs/how-to/hardening-your-cluster.

[4] https://aws.github.io/aws-eks-best-practices/security/docs.

[5] https://learn.microsoft.com/en-us/azure/aks/operator-best-practices-cluster-security?tabs=azure-cli.

are tricky for users and lead to misconfiguration that may endanger data manipulated by these containers. Moreover, we saw that security policy can differ from one Cloud provider to another, leading to a complex matrix of security configuration depending on the selected provider, even more so in a multi-Cloud context.

## 3  Contribution Proposal

ABAC is far from being a self contained product. It relies on top of other services that run the organization infrastructure. The core of ABAC is the PDP and access control rules database. ABAC core retrieves attributes through its Policy Information Point (PIP). PIP is a bridge between PDP and attributes database. Attributes can be stored in files, a DataBase Management System (DBMS) or in a LDAP (Lightweight Directory Access Protocol) directory. We have two types of attributes: user's attributes and object's attributes. Both can be stored on the same database, or they can be split. For example, you can have user's attributes in the Identity Access Management (IAM) and object's attributes in a standalone database.

Each application willing to support ABAC attributes credentials must have a PEP module. This module handles the user's access request and sends it to the PDP. There is no consensus on the information sent by PEP to PDP. In the NIST report [11], Fig. 4, PEP only sends the object's and the user's identifier. PDP asks PIP to lookup for attributes to perform access control. However in the open source implementation Cerbos[6], the PEP module performs the lookup and sends the attributes straight to the PDP. Casdoor's[7] implementation is closer to NIST preconisations. ABAC lacks consensus regarding implementation design.

Our contribution proposal is to use ABAC as a backend to increase storage services security without overloading users with technical details. Our idea is that users understand the meaning of their attributes. For instance, if a doctor has *Medecine* and *France* attributes, he understands that his data have a *Medecine* level of confidentiality and should stay in *France*. However, he is not able to write a deployment policy for Cloud orchestrator that matches his constraints. Moreover, configuration can change from one Cloud provider to another. We need an intermediate service between users and resources that:

1. Has its own PEP module to authenticate users;
2. Retrieves user's attributes from ABAC databases and eventually map them to the provider;
3. Has a repository of scenarios to satisfy users requirements. A scenario could be an Ansible recipe, a Kubernetes deployment, a simple shell script etc. It will talk with the storage provider to set up the storage facility required by the user.

---

[6] https://cerbos.dev/.
[7] https://casdoor.org.

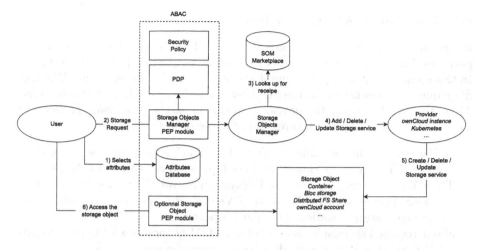

**Fig. 4.** Storage Objects Manager (SOM)

Figure 4 shows the workflow of the Storage Objects Manager (SOM). First, the user selects some attributes of its own (1). Then, he submits a request for a Storage Object (SO) labeled with selected attributes to the SOM (2). A SO is a storage resource such as a database, an account on a web drive service, a dedicated web drive service, an iSCSI target etc. SOM looks up in its marketplace (3). The marketplace contains recipes to manage SO. Then, the SOM uses a selected recipe to ask the provider to create the SO (4). The provider can be a Kubernetes orchestrator, a web drive service or an on the shelf Storage Area Network (SAN) device etc. In the end, the provider instantiates the SO (5). When the SO is available, the requester can use it (6). The SO itself can also embed its own PEP module, making it possible to use ABAC authentication for fine grained access control. We can notice that with our approach, even if the SO does not implement its own PEP module, attributes provide security to underlying layers.

## 4    General Use-Cases

In this section we will discuss two general uses-cases. The first one is a creation request of a web drive service. This service will run in a container managed by a Kubernetes orchestrator. This container will use a Persistent Volume (PV) as the storage backend. We assume that the web drive service does not implement a PEP module. Consequently, this use case will only rely on the user's attributes. The second use-case will present a request for an account on a web drive service. We will assume that this web drive service implements a PEP module. For the first use-case, we will consider a user with a single *France* attribute. In the second use-case, the user has *Medecine* and *France* attributes.

### 4.1   Use Case 1: ABAC User Profile

In this use-case, the user must create a web drive service to share data with partners. For data sovereignty sake, data must be physically located in France. In this scenario, the user's SO request is tagged with *France* attribute. When the SO request reaches the SOM, the SOM retrieves the recipe from its marketplace. Then, the SOM contacts the provider. In this case, the provider is Kubernetes. The user's request implies the creation of three objects:

1. A Persistent Volume Claim (PVC). The PVC is a claim for a Persistent Volume. The PV is the lower level configuration of storage for Kubernetes. The PV can represent a Network Filesystem Server (NFS), iSCSI, or a cloud-provider-specific storage system etc. PVs embed labels that give information on storage nature, location, physical properties etc. A PVC is a user's wish about required storage for its container. SOM generates a PVC that requires a PV located in France according to attribute *France*;
2. A container that hosts the web drive service. Kubernetes runs its containers on virtual or physical nodes. Each node embeds a container engine that performs the instantiation of the container. The attribute *France* means that container must run on a node located in France;
3. A regular web drive service. This service does not embed a PEP module. Consequently, it can not use attributes for authentication purposes.

SOM generates a regular Kubernetes deployment for these two objects according to constraints implied by attributes. Then, it sends it to the Kubernetes provider. Finally, Kubernetes creates our SO and informs the SOM. In this use-case, we can see that even if the web drive service is regular (i.e., does not come with a PEP module), we can apply some security measures regarding SO physical location.

### 4.2   Use Case 2: ABAC Authentication

In this use-case, the user is a doctor that requests an account on a web drive service for medical purposes. This time, the user has *Medecine* and *France*. As stated in the first use-case, attribute *France* means that PV and container must be located in France. The *Medecine* attribute adds some constraints:

– The datacenter that runs the web drive service must be at least Tier3 regarding power supply, redundancy, physical access security etc.;
– As stated before, containers runs on top of an engine. There are several container engines. Some of them are general purpose like Docker or CRI-O, some of them are hardened like gVisor or Kata Containers. As with PV, nodes have labels, especially information about available container engines. User has *Medecine* attribute, it means that data are confidential. So, a security issue in the web drive service may have serious consequences on confidentiality. The SOM will interpret this attribute as a directive to choose a web drive service that runs on a hardened container engine.

The *Medecine* attribute also implies that the SOM must choose a web drive service ABAC capable. Indeed, users with *Medecine* attribute must not be able to create public shares for their data. Moreover, they should only share data with others *Medecine* labeled users. To enforce these security measures, the web drive service must be able to deal with ABAC attributes (see Fig. 5).

## 5    Geom-SAC Use-Case and ABC Preliminary Model

Dr. Smith is a Lead Researcher at a pharmaceutical research organization cooperating with Geom-SAC team. He is responsible for overseeing a critical research project, Drug+ Project, which involves the development of a groundbreaking drug candidate using Geom-SAC. Dr. Smith's user attributes play a significant role in determining his access to sensitive molecular data and his actions within the Geom-SAC research platform as it is discussed in the following:

- His Access Control:
  - **Role-Based Access:** Dr. Smith's role as a Lead Researcher grants him elevated access privileges within the organization's molecular Geom-SAC research platform. He can create, modify, and delete molecular data as needed for the Drug+ Project.
  - **Project Affiliation:** Being affiliated with Drug+ Project allows Dr. Smith to access all project-specific data, including confidential research findings, molecular structures, and experimental results produced by GEO-SAC.
  - **Sensitivity Level:** Dr. Smith is granted access to high-sensitivity data, which includes proprietary formulas and research findings that are crucial to the success of the Drug+ Project.
  - **Credentials:** Dr. Smith's Ph.D. in Molecular Chemistry certifies his qualifications to handle and make decisions about the Drug+ project's molecules.
- His Actions:
  - **Data Modification:** Dr. Smith has the authority to modify molecular data and experiment parameters to optimize the drug candidate within the Drug+ Project.
  - **Data Access:** He can access and review all sensitive molecular data related to the Drug+ Project, allowing him to monitor progress and make informed decisions.
  - **Collaboration:** As a Lead Researcher, he can collaborate with other team members, including researchers from GEO-SAC, data analysts, and administrators, to drive the Drug+ project forward.
- His Responsibilities:
  - **Drug+ Project Oversight:** Dr. Smith is responsible for overseeing the entire Drug+ research project, ensuring that the project meets its goals and stays on schedule.
  - **Data Security:** He must maintain the security and confidentiality of high-sensitivity data, ensuring that it is not accessed by unauthorized users.

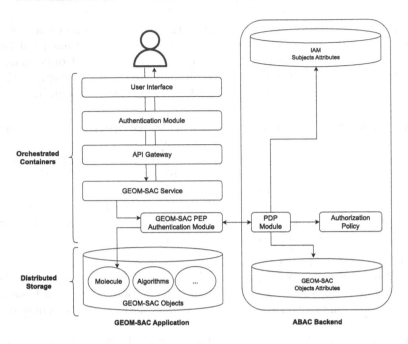

**Fig. 5.** ABAC implementation Geom-SAC

Each category in this set have several sets of functionalities, roles, contribution levels for the project. This set of users of Dr. Smith's lab consists of three categories (Researchers, Admins, Directors)

1. **Academic researchers:** This set of users (e.g. students, junior researchers, senior researchers) are benefiting from many privileges for free that aid them in their research, but with limitations according to their seniority in research. The properties of this set can be described as follows:
   – **Access control:**
      • **Role-Based Access:** In this set, users have rights to create, modify, or delete molecular data. Also, they have limited access to CPU hours. Can be upgraded according to their seniority in research.
      • **Project Affiliation:** Ensures each researcher can access and contribute only to specific projects he is assigned to, as it grantees access to highly sensitive data, includes molecular formulas, and research findings.
      • **Sensitivity levels:** Researchers will have access to specific set of molecules, based on molecular type (Proteins, Polymers, Lipids, etc.). For the free version, they will have access to molecular databases with small molecules of maximum 9 atoms.
      • **Credentials:** each researcher must authenticate using his own credentials (user-name and password) to access data-sets related to molecular properties.

- **Actions:**
  - **Targeted molecular generation:** The researcher can generate novel molecules similar to a specific reference molecule with limited number of molecules per day for free. can be upgraded.
  - **Molecular data storage:** Researchers can store molecules up to certain space for free. can be upgraded.
  - **Geom-SAC model modifications:** Researchers an view some of Geom-SAC model parameters,and modify for better customization. This includes modifying number of layers, number of neurons, and graph encoder model.
  - **Collaboration:** Researchers can share their findings with colleges and the director of the lab (as Dr. Smith) within the same research lab, even with different project affiliations.
- **Responsibilities:**
  - **Running Experiments:** Responsible for designing and conducting the process of molecular generation and optimization, and storing the output molecules based on their attributes.
  - **Optimizing resources:** Ensures the experiment is done while optimizing the resources (e.g.: terminate the CPU while not in use to reduce money costs).

2. **Administrators:** Administrators play a crucial role in managing and maintaining the project. Their responsibilities include monitoring the system, assigning/removing user roles, ensuring security, and managing the overall functionality of the system. Their responsibilities can be defined as follows:
   - **Access control:**
     - **Role-Based Access:** Administrators define and manage user roles in Geom-SAC, such as Director, Academic Researcher, and Administrator, assigning specific access permissions based on responsibilities.
     - **Project Affiliation:** Administrators manage project access by assigning users to specific projects, ensuring that they have access only to data and functionalities related to their affiliated projects by mapping researchers attributes with molecular data attributes.
     - **Sensitivity levels:** Administrators define sensitivity levels for molecular data in the Geom-SAC project, categorizing data as low, medium, or high sensitivity, and implement access control policies to restrict access and modifications among users based on these sensitivity levels.
     - **Credentials:** Administrators control user authentication by managing the setup of login credentials, including enforcing strong password policies and handling requests for credential updates, wherein they verify users' identities before making any changes to their roles or when requesting new permissions.
   - **Actions:**
     - **Define and Manage User Roles:** Establish and monitor roles for each user to ensure appropriate access levels for each of them.
     - **Manage Authentications and Credential Updates:** Supervise the setup and maintenance of user authentication, including strong

password policies, and handle requests for credential updates with identity verification.

- **Enforce Access Control Policies:** Establish access control policies based on user roles, project affiliations, and sensitivity levels to ensure secure and controlled access.

    – **Responsibilities:**
    - **Policy Setup:** Defining and monitoring policies considering roles, affiliations, and sensitivity levels.
    - **Attribute-Based Role Assignments:** Assess user attributes, and dynamically adjust roles based on changes to align their responsibilities with the desired output of the project.

3. **Board of Directors:** A board of Directors is responsible for providing insights of the project and strategic planing to ensure the success of the research project, as well as allocating resources for researchers, engaging with the key stakeholders, and risk management to ensure efficiency and effectiveness of the project.

    – **Attribute-Based Strategic Oversight:**
    - **Evaluate Project's Attributes:** assess project attributes, such as project affiliations, sensitivity levels, and overall goals, to ensure alignment with organizational goals.
    - **Strategic Alignment Analysis:** Conduct a strategic analysis of the attributes to ensure the alignment with the organizational strategy and research priorities.
    - **Resource Allocation Decision:** Based on attribute assessments of researchers and molecular data, The board of directors makes decisions regarding resource allocation for the project, optimizing efficiency and impact.

    – **Review and Approval of Access Requests:**
    - **Attribute-Based Access Evaluation:** The board of Directors evaluates access requests considering attributes such as user roles, project affiliations, data attributes and sensitivity levels.
    - **Policy Adherence Check:** Ensuring access requests adhere to established access control policies and align with project objectives.
    - **Risk Assessment:** Conduct risk assessments based on attribute analysis to reduce potential security risks.

    – **Attribute-Based Monitoring:**
    - **Attribute-Based Performance Control:** Generating reports includes attribute-based performance metrics, evaluating factors such as project progress, and collaboration effectiveness.
    - **Dynamic Improvement:** Leveraging attribute-based insights from reports to identify areas of improvement within the project.

In this Geom-SAC use case, Dr. Smith's user attributes, including his role, project affiliation, sensitivity level, and credentials, grant him the necessary access and permissions to effectively lead and manage Project Drug+. These attributes help in defining his role, responsibilities, and actions within the Geom-SAC research platform, while also ensuring the security and integrity of sensitive data.

# 6    Conclusion

ABAC is the next generation authorization model for dynamic environments such as Cloud infrastructures. Major Cloud providers progressively make ABAC available for their customers. The use of attributes evaluation for authorization avoids administrators to write new access rules when a new object or subject appears. This makes ABAC more scalable than IBAC or RBAC. Moreover, with an extra cost of development, it is possible to make application ABAC aware to provide a very fine grained access control on its objects. As it is a relatively new access control model, ABAC lacks implementation consensus. It makes it difficult to implement an ABAC backend.

In this position paper, we proposed to make ABAC attributes an asset to empower users with their security. We considered the meaning of attributes to translate them in terms of security properties on services deployments. Indeed, service deployment configuration on a Cloud infrastructure can be cumbersome for regular users. Each Cloud provider supplies its own security features as well as its own set of resource labels to pinpoint them. Our contribution is to set a mapping between ABAC attributes (information that users understand) and deployment configuration (that requires knowledge on orchestrator configuration). This enables users to regain control of their data security. Moreover, if the application is ABAC aware, it leads to an end-to-end security policy from the deployment to the access to objects.

# 7    Future Works

Our next step is to implement a proof of concept for our approach. We will set up an ABAC backend in our Cloud infrastructure. Then, we will work in parallel on 1) the mapping between attributes and security properties regarding deployments in our Kubernetes orchestrator and 2) The instrumentation of Geom-SAC application with a PEP module to enable fine grained access to the application's attributes such as molecule or algorithms. Our goal is to increase security of GEAOM-SAC from the deployment to the access control on objects. We will evaluate the security of this implementation and we will also perform some benchmarking to measure fine grained access control impact on performances.

**Acknowledgment.** We extend our heartfelt gratitude to the National Institute of Informatics (NII) and the Centre National de la Recherche Scientifique (CNRS) for their invaluable collaboration through the signed MOU, which has paved the way for this impactful joint research.

**Funding Information.** This work is funded by the French Ministry of Higher Education and Research in the context of DATACENTER 2023. This is a governmental call for projects to provide a secure deployment of storage services in the Cloud. We would like to thank you the National Institute of Informatics (Japan) for its support for this research.

# References

1. National Security Agency. Kubernetes hardening guidance v 1.2. Technical report National Security Agency, Cybersecurity and Infrastructure Security Agency (2022). https://media.defense.gov/2022/Aug/29/2003066362/-%201/-%201/0/CTR_KUBERNETES_HARDENING_GUIDANCE_1.2_20220829.PDF

2. Badger, L., et al.: Practical domain and type enforcement for UNIX. In: Proceedings 1995 IEEE Symposium on Security and Privacy, pp. 66–77. IEEE (1995)

3. Bell, D.E.,: Secure computer systems: mathematical foundations and model. In: Technical Report ESD-TR-73-278-1 (1973)

4. Biba, K.: Integrity considerations for secure computing systems. In: Mitre Report MTR-3153, Mitre Corporation, Bedford, MA (1975)

5. Boebert, W.E.: A practical alternative to hierarchical integrity policies. In: Proceedings of the 8th National Computer Security Conference, 1985 (1985)

6. Bousquet, A., et al.: Enforcing security and assurance properties in cloud environment. In: 2015 IEEE/ACM 8th International Conference on Utility and Cloud Computing (UCC), pp. 271–280. IEEE (2015)

7. Brand, S.L.: DoD 5200.28-STD department of defense trusted computer system evaluation criteria (orange book). Nat. Comput. Secur. Center, 1–94 (1985)

8. Briffaut, J., Lalande, J.-F., Toinard, C.: Formalization of security properties: enforcement for mac operating systems and verification of dynamic mac policies. Int. J. Adv. Secur. 2(4), 325–343 (2009)

9. Briffaut, J., Lalande, J.-F., Toinard, C.: Security and results of a large-scale high-interaction honeypot. J. Comput. 4(5), 395–404 (2009)

10. Briffaut, J., et al.: Enforcement of security properties for dynamic MAC policies. In: 2009 Third International Conference on Emerging Security Information, Systems and Technologies, pp. 114–120. IEEE (2009)

11. Chung et al.: Guide to Attribute Based Access Control (ABAC) Definition and Considerations (2019). https://doi.org/10.6028/NIST.SP.800-162

12. Cuppens, F., Cuppens-Boulahia, N.: Modeling contextual security policies. Int. J. Inf. Secur. 7(4), 285–305 (2008)

13. Denning, P.J.: Fault tolerant operating systems. ACM Comput. Surv. (CSUR) 8(4), 359–389 (1976)

14. Downs, D.D., et al.: Issues in discretionary access control. In: 1985 IEEE Symposium on Security and Privacy, pp. 208–208. IEEE (1985)

15. Ennahbaoui, M., Elhajji, S.: Study of access control models. In: Proceedings of the World Congress on Engineering, Vol. 2, pp. 3–5 (2013)

16. Ferraiolo, D., Cugini, J., Kuhn, D.R., et al.: Role-based access control (RBAC): features and motivations. In: Proceedings of 11th Annual Computer Security Application Conference, pp. 241–48 (1995)

17. Ferraiolo, D.F., Barkley, J.F., Kuhn, D.R.: A role-based access control model and reference implementation within a corporate intranet. ACM Trans. Inf. Syst. Secur. (TISSEC) 2(1), 34–64 (1999)

18. Ferrario, D.: Role-based access control. In: Proceedings of 15th National Computer Security Conference, 1992 (1992)

19. Gros, D., et al.: PIGA-cluster: a distributed architecture integrating a shared and resilient reference monitor to enforce mandatory access control in the HPC environment. In: 2013 International Conference on High Performance Computing & Simulation (HPCS), pp. 273–280. IEEE (2013)

20. Gupta, E., et al.: Attribute-based access control for NoSQL databases. In: Proceedings of the Eleventh ACM Conference on Data and Application Security and Privacy, pp. 317–319 (2021)
21. Harrison, M.A., Ruzzo, W.L., Ullman, J.D.: Protection in operating systems. Commun. ACM **19**(8), 461–471 (1976)
22. ITSEC. Information Technology Security Evaluation Criteria (1991)
23. Jahid, S., et al.: MyABDAC: compiling XACML policies for attribute based database access control. In: Proceedings of the First ACM Conference on Data and Application Security and Privacy, pp. 97–108 (2011)
24. Kubernetes – see. https://kubernetes.io/.k8s
25. Lampson, B.W.: Protection. ACM SIGOPS Oper. Syst. Rev. **8**(1), 18–24 (1974)
26. Li, N., Mao, Z.: Administration in role-based access control. In: Proceedings of the 2nd ACM symposium on Information, Computer and Communications Security, pp. 127–138 (2007)
27. Loscocco, P.A., et al.: The inevitability of failure: the flawed assumption of security in modern computing environments. In: Proceedings of the 21st National Information Systems Security Conference (1998)
28. Mahmoud, A.A., et al.: A new graph-based reinforcement learning environment for targeted molecular generation and optimization. In: 2023 12th International Conference on Software and Information Engineering (2023)
29. Mayfield, T., et al.: Integrity in automated information systems. Nat. Secur. Agency, Tech. Rep. **79** (1991)
30. Osborn, S., Sandhu, R., Munawer, Q.: Configuring role-based access control to enforce mandatory and discretionary access control policies. ACM Trans. Inf. Syst. Secur. (TISSEC) **3**(2), 85–106 (2000)
31. Rahman, M.U.: Scalable role-based access control using the EOS blockchain. arXiv preprint: arXiv:2007.02163 (2020)
32. Sandhu, R., et al.: The NIST model for role-based access control: towards a unified standard. In: ACM Workshop on Role-Based Access Control, vol. 10, no. 344287.344301 (2000)
33. Sandhu, R., Munawer, Q.: How to do discretionary access control using roles. In: Proceedings of the Third ACM Workshop on Role-Based Access Control, pp. 47–54 (1998)
34. Sandhu, R.S.: The typed access matrix model. In: IEEE Symposium on Security and Privacy, pp. 122–136. Citeseer (1992)
35. Sandhu, R.S.: The schematic protection model: its definition and analysis for acyclic attenuating schemes. J. ACM (JACM) **35**(2), 404–432 (1988)
36. Sultan, S., Ahmad, I., Dimitriou, T.: Container security: issues, challenges, and the road ahead. IEEE Access **7**, 52976–52996 (2019)
37. Wang, X., Du, J., Liu, H.: Performance and isolation analysis of RunC, gVisor and Kata containers runtimes. Cluster Comput. **25**(2), 1497–1513 (2022)
38. Zhang, Z., Zhang, X., Sandhu, R.: ROBAC: scalable role and organization based access control models. In: 2006 International Conference on Collaborative Computing: Networking, Applications and Worksharing, pp. 1–9. IEEE (2006)

# Machine Learning to Model the Risk of Alteration of Historical Buildings

Baptiste Petit[1], Emilie Huby[2], Céline Schneider[1], Patricia Vazquez[1], Cyril Rabat[1], and Hacène Fouchal[1(✉)]

[1] Université de Reims Champagne-Ardenne, Reims, France
hacene.fouchal@univ-reims.fr
[2] Laboratoire de Recherche des Monuments Historiques (LRMH), CRC – MNHN, CNRS, Ministère de la Culture – UAR 3224, Champs-sur-Marne, France

**Abstract.** Detaild, realistic knowledge of the weather effects on building materials is essential to efficiently maintain and manage historical buildings. The main aim of this study is to model the long-term weathering behavior of stone materials under different climate change scenarios. In order to achieve this, data have been collected from an adapted IoT architecture on an emblematic monument in the city of Reims, the "Basilique St-Remi". An Adhoc wireless sensor network has been deployed during two years. These sensors have been located in various locations of the building walls.

A large amount of data has been analyzed together with general weather data of Météo France. Some features about the data variations have been extracted and split into different clusters. The behavior of stones regarding humidity and temperature has been modeled. The last step was the prediction of the behavior of the whole building in the future (in the next 50 or 100 years) according to the weather expectations in various scenarios given by climate changes.

This work is a first attempt to assess precise behavior of historical buildings and gives enough information to decision makers to choose relevant measures to consider for building preservation.

This study has used around 1400 mega bytes of inputs and has considered three possible scenarios for climate change. The prediction process has been computed on a personal computer during few minutes.

The obtained results are precise enough compared to general predictions studied in the past.

**Keywords:** Deep learning · Data collection · Machine Learning · Building preservation · Weather prediction

## 1 Introduction

The assessment of stone alteration of historic buildings is essential in order for better maintenance and conservation of the cultural heritage. The stone weathering is a complex phenomenon, where different factors come into play such as

É. Renault et al. (Eds.): MLN 2023, LNCS 14525, pp. 218–231, 2024.
https://doi.org/10.1007/978-3-031-59933-0_15

external factors (climatic conditions, exposure, human activities) and internal factors (stone properties) [1].

A common approach to study stone weathering, and in particular compare different stones weathering behaviors, is through accelerated aging tests. These tests subject the samples to harsh conditions to reproduce specific weathering mechanisms (thermal shock, freeze/thaw cycles, salt weathering), causing significant damage allowing to compare the stones' resistance. Because of their harsh conditions, these tests are not representative of the actual weathering conditions on a monument and can lead to erroneous conclusions [2,3] They allow assessing the response of materials to realistic strains . Other experimental tests are thus developed, reproducing the environmental variations and their geometry in a more realistic manner [4–7]. The downside of such tests is that, in order to assess the long-term behavior, many cycles have to be carried out, which is time consuming.

The novel approach proposed here is to associate data obtained through laboratory realistic tests with machine learning in order to estimate the stone alteration on the long-term.

The rest of this paper is structured as follows: Sect. 2 presents the methodology of data collection. Section 3 presents briefly the theoretical model of stone behaviour. Section 4 is dedicated to the analysis of the collected data and Sect. 5 presents concludes the study.

## 2    Data Collection

### 2.1    Parameters

The parameters chosen for this study are temperature and relative humidity variations. Among environmental conditions, temperature and humidity are indeed particularly important for the stone weathering. For most stones, temperature and humidity variations can induce stress due to thermal and hygric expansions/contraction and their repetition can potentially induce stone failure [8]. The temperature also influences other factors such as moisture availability or evaporation rates [9]. Humidity variations intervene in different weathering mechanisms for wind- and rain-exposed areas and sheltered walls [10,11]. Humidity and water accessibility are also essential factors for biological colonization.

### 2.2    Case Study

The historical building studied here is the Saint-Remi Basilica, an emblematic monument in Reims. This historical building was built during the 11th century and is listed UNESCO World Heritage site. This monument is representative of the regional built heritage. The monument is composed of different materials, among them the Courville and the Savonnières limestones. The Courville limestone is a local stone and considered the original stone of the Saint-Remi Basilica. This limestone has been used for buildings in the area around Reims since the medieval period. The Savonnières limestone is a regional stone used in Reims and its surrounding since the middle of the 19$^{th}$ for construction and restoration purposes.

A previous study allowed to study the environment of the basilica in order to identify typical environmental variations [12]. The access to the monument allowed to implement the sensors' network on the two towers of the main façade (Fig. 1).

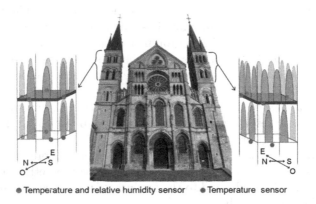

● Temperature and relative humidity sensor    ● Temperature  sensor

**Fig. 1.** Schematic representation of the sensors network on the Saint-Remi basilica

## 2.3    On-Site Monitoring

A sensors' network has been implemented on the monument for 2 years, collecting temperature and relative humidity variations every hour. The sensors used here were i-buttons, small and reliable sensors which are appropriate for an *in situ* monitoring on a historical building. The hourly values obtained for one sensor are represented in Fig. 2.

The analysis of the monitoring data for the whole sensors' network lead to the identification of micro-climates on the monument and typical events. These events, hereafter called typical days, constituted of repetitive temperature and relative humidity variations. The characterization of the typical days is the following:

- *Sunny day*: high temperature (temperature range > 10 °C), low relative humidity (RH < 90%)
- *Cold Rainy day*: low temperature (temperature range < 10 °C), high relative humidity (RH > 90%)
- *Warm Rainy day*: high temperature (temperature range > 10 °C), high relative humidity (RH > 90%)
- *Frost day*: low temperature (minimal temperature < 0 °C), high relative humidity (RH > 90%)

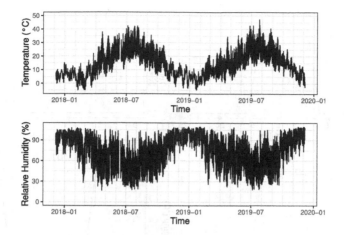

**Fig. 2.** Temperature and relative humidity variations of a sensor (South tower - 30 m high - East oriented) in 2018 and 2019 [12]

## 3    Modeling

### 3.1    Model

The categorization of the typical days mentioned in Sect. 2.3 was used to model the temperature and relative humidity variations. For each typical days, the variations were modeled as follows:

- Sunny day (Fig. 3a) A sunny day cycle represented a temperature increase during the day. It began with 6 h at 20 °C. The temperature then increased up to 35 °C with a 0.125 °C/min rate and remained at 35 °C for 10 h. Finally, the temperature decreased down to 20 °C with a −0.125 °C/min rate and remained at 20 °C for 8 h.
- Cold rainy day (Fig. 3b) A cold rainy day cycle represented a day with a rain event and no particular temperature variation. The temperature thus remained stable at 20 °C during the whole cycle. A 1 h capillary imbibition occurred 5 h after the beginning of the cycle.
- Warm rainy day (Fig. 3c) A warm rainy day cycle represented a day with a rain event and a temperature increase during the day. It began with 5 h at 20 °C. During the 6th hour, a capillary imbibition occurred. After this simulated rain event, the temperature increased up to 35 °C with a 0.125 °C/min rate and remained at 35 °C for 10 h. The temperature then decreased down to 20 °C with a −0.125 °C/min rate and remained at 20 °C for 8 h.
- Frost day (Fig. 3d) A frost day cycle represented a day with a very low temperature (under 0 °C). Before the cycle, a 1 h capillary imbibition took place. The temperature decreased to 5 °C and the cycle began after stabilization. The temperature was maintained at 5 °C for 5 h. The temperature then decreased down to −10 °C at a −0.125 °C/min rate and at −10 °C for 10 h. The temperature increased back to 5 °C and remained at 5 °C for 8 h.

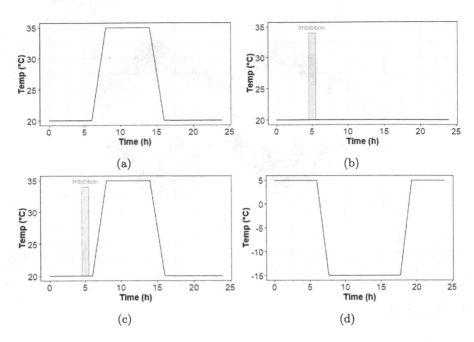

**Fig. 3.** Modeling of the typical days: Sunny (a), Cold Rainy (b), Warm Rainy (c) and Frost (d) days

## 3.2  Experimental Setup

Experimental realistic tests were developed [7], aiming to reproduce the typical environmental variations previously identified. The experimental setup is presented in Fig. 4 and consisted of:

- the stone sample, representing a vertical stone on the monument
- a programmable heating/cooling plate, simulating temperature variations on the exposed surface
- a water container, simulating rain events through a 1-hour capillary imbibition on the exposed surface

The materials studied were the Courville and the Savonnières limestones, mentioned in 2.2. The sample's behavior is analyzed throughout the experiment thanks to thermo-couples and strain gauges (Fig. 5). Strain gauges were placed on the surface at different distances from the cooling/heating plate (0.5, 1.3 and 2.6 cm) and at different heights (5, 6, 7 and 8 cm). The placement of the thermo-couples in the sample corresponded to the placement of the strain gauges (distance from the plate, height). The collected data consisted of the temperature and strain behavior of the stone during 5 cycles reproducing the identified typical days.

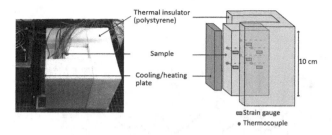

**Fig. 4.** Experimental setup

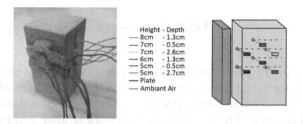

**Fig. 5.** Sample: setting, showing the location of thermocoupes (height and depth) and strain gauges

## 4    Data Analysis

Based on data obtained from laboratory experiments and meteorological data from various sources, our study aims to predict the stress suffered by the stones of the Basilica Saint-Remi in Reims caused by temperature and humidity variations. This behavioral prediction will allow us to predict the stress of the stones composing the entire building over the next 50 or 100 years, according to 3 climate change scenarios.

### 4.1    Temperature Propagation

By analyzing the propagation process, it becomes possible to predict with greater accuracy the stress undergone by the rock at different depths. This prediction of thermal propagation is a crucial step in assessing the rock's response to temperature variations. The calculation of heat propagation is based on data from the laboratory experiments presented previously. These data include the surface temperature of the rock, the ambient temperature of the room, and temperature readings at 0.3, 1.3 and 2.6 cm depth. The graph in Fig. 6 shows the temperatures recorded during a Sunny Day cycle on a Courville stone.

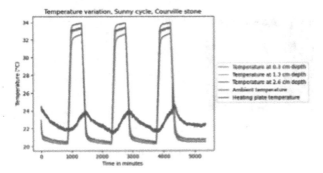

**Fig. 6.** Temperature

In order to predict the temperature propagation within the rock, we use a neural network for each typical day (sunny, ) and for each type of rock. For the training stage, we consider each cycle in its entirety to maintain the temporal consistency of the data. The input for each network is composed by the following parameters: the temperature of the heating plate, the ambient temperature, the humidity of the rock and the specific depth for which we want to predict the temperature. The output is the temperature corresponding to the depth specified in the input. In Fig. 7, the architecture of the neural networks used is presented.

```
Model: "sequential_4"

_____
Layer (type)                 Output Shape              Param #
=================================================================
simple_rnn_4 (SimpleRNN)     (None, None, 256)         67872

dense_16 (Dense)             (None, None, 128)         32896

dense_17 (Dense)             (None, None, 128)         16512

dense_18 (Dense)             (None, None, 64)          8256

dense_19 (Dense)             (None, None, 1)           65

=================================================================
Total params: 124,801
Trainable params: 124,801
Non-trainable params: 0
_____
```

**Fig. 7.** Neural Network

In the Fig. 8, the temperature of the rock at 0.3cm depth (shown in blue) and the temperature predicted by the neural network at the same depth are represented. These data relate to a Courville stone during a Sunny Day. The predicted data show good correlation with the experimental data, suggesting the adequacy of the model.

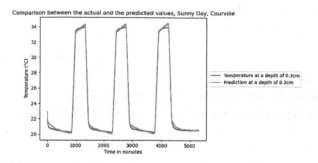

**Fig. 8.** Depth

## 4.2 Stress

Once propagation has been predicted, a second set of neural networks is trained to predict stress as a function of temperature. To obtain the stress experienced by the rocks, we use the Eq. 1, where $\sigma$ is the stress, $\epsilon$ is the strain and $E$ is the modulus of elasticity of the rock (Young modulus).

$$\sigma = E * \epsilon \tag{1}$$

To obtain the deformation $\epsilon$, we use the Eq. 2, with $R$ the measured strength value, $R_{initial}$ the average strength over the first 20 min, $T$ the measured temperature value, $T_{initial}$ the average temperature over the first 20 min. $GF$ and $\alpha_{gauge}$ are the gauge factor and the linear thermal expansion coefficient, characteristics of the strain gauge (Kyowa strain gauge KFWS-2N-120-C1-11 L3M2R); their values are taken from the technical sheet and are respectively 2.12 and $11.6E^{-6}$ °C$^{-1}$.

$$\epsilon = \frac{\Delta L}{L} = \frac{((R - R_{initial})/R_{initial})}{GF} + \alpha_{gauge} * (T - T_{initial}) \tag{2}$$

The mechanical properties of the stones were measured in laboratory experiments [13] (Table 1). The values for the modulus of elasticity are used for the calculation in Eq. 1. The calculated stress in the rock can be compared with the rock's tensile and compressive strength values. If the stress is greater than the tension threshold, cracks will appear, and if the stress is greater than 40% of the threshold, the surface of the rock will loosen. It is interesting to note that the mechanical properties in the wet conditions (48h immersion in water) are lower than in the dry conditions. For instance, the tensile strength of the Courville limestone is reduced by 70% and the compressive strength by 35%.

**Table 1.** Mechanical properties for the Courville and Savonnières limestones measured on samples (ø = 4 cm, h = 8 cm)

|  | Courville | Savonnières |
|---|---|---|
| Young Modulus (GPa) | 3.8 | 1.8 |
| Compressive strength (MPa) : dry/wet | 27/17 | 11/7 |
| Splitting tensile strength (MPa) : dry/wet | 3.0/0.9 | 1.8/0.8 |

Below is a graph showing the stress undergone by a Courville stone during a Frost Day cycle (Fig. 9). The stress is calculated at a depth of 1.3 cm. The loosening and cracking thresholds are also plotted. In this case, the stress induced by a Frost day does not seem to induce damage in the stone.

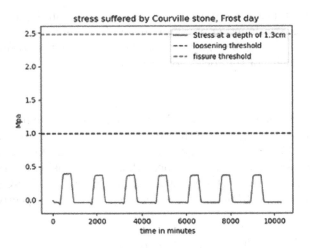

**Fig. 9.** Calculated stress generated in Courville limestone during a Frost day

Once the constraint has been calculated, the neural networks can be trained. Each of the neural networks is trained on a specific type of day, a specific type of rock and a specific depth. As an input, the network uses the temperature at the depth where we want to predict the stress, and as an output, we retrieve the stress at the specified depth. The architecture of the neural network is presented in Fig. 10.

After observing a drop in the accuracy of the network in predicting negative stresses, we decided to apply a coefficient to the values. First, the minimum observed constraint was recovered, then its absolute value was added to all the values. This operation transforms the negative values into positive values while preserving the consistency of the data. Once the network has been trained, this coefficient was substracted from the predicted values to obtain the original values. The stress values experienced by the rock and the stress values predicted by

```
Model: "sequential_1"

Layer (type)              Output Shape              Param #
=====================================================================
simple_rnn_1 (SimpleRNN)  (None, 1024)              1050624

dense_4 (Dense)           (None, 1024)              1049600

dense_5 (Dense)           (None, 1024)              1049600

dense_6 (Dense)           (None, 1024)              1049600

dense_7 (Dense)           (None, 1)                 1025

=====================================================================
Total params: 4,200,449
Trainable params: 4,200,449
Non-trainable params: 0
```

**Fig. 10.** Neural Network..

the neural network are illustrated in Fig. 11. The average error of the network, after training, is 0.001.

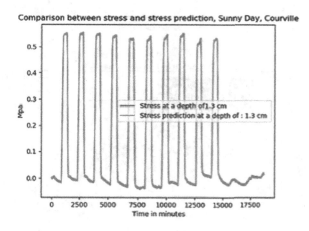

**Fig. 11.** Comparison of calculated stress and stress prediction for a Courville limestone during a Sunny day

## 4.3   France Meteo Data

In order to verify the data observed at the Saint-Remi Basilica, the weather data from Météo France for the same period was retrieved. These data were obtained from observations of international surface observation messages (SYNOP) circulating on the World Meteorological Organisation's (WMO) global telecommunications system (GTS). It should be noted that these meteorological data were collected at a measuring point located approximately 9km from the Saint-Remi Basilica.

The weather data, average daily temperature and humidity, obtained by Météo France and measured on the Basilica over the period from 5 December 2017 to 27 February 2020 are represented in Fig. 12. The weather data from Météo France is consistent with the observations made at the Basilica described in Sect. 2.3. There are a few differences, which can be explained by the accuracy of the sensors at the Basilica and the distance between the two measurement points.

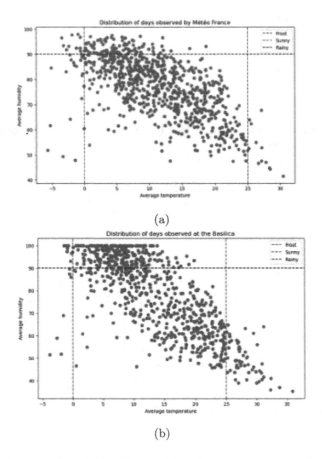

(a)

(b)

**Fig. 12.** Weather data from Météo France (a) and measurements on the Basilica (b)

## 4.4    RIAS Data

To assess future risks, it is essential to estimate future weather conditions. This estimate is based on DRIAS data. DRIAS provides climate projections developed by the IPSL, CERFACS and CNRM climate modelling laboratories. The projection data was obtained for a location near the Saint-Remi Basilica. The three global warming scenarios concern a future with non reduced emissions, moderate emissions and controlled emissions. These data take the following form:

- Scenario: indicates the global warming scenario envisaged.
- Horizon: representing the time interval considered
- Month: designates the month concerned
- Maximum, minimum and average temperature: details the maximum, minimum and average temperature values.
- Number of hot days: indicates the number of hot days.
- Number of days with frost: indicates the number of days with frost.
- Number of days with rain: indicates the number of days with rain.

### 4.5   Data Augmentation

The collected data cover a considerable period of time, while the network is trained over shorter periods. Consequently, it is necessary to increase the data, focusing on temperature and humidity information. To do this, we rely on the forecasts issued by DRIAS, and by combining these long-range forecasts with short-range observations from Météo France. Pairs of temperature and humidity values for each day over a period of one hundred years are thus generated. The obtained result for the average humidity and temperature for days in 2071 is represented in Fig. 13.

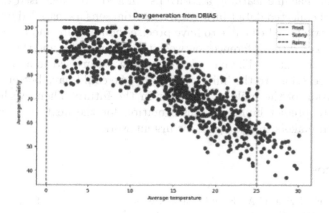

**Fig. 13.** Generated daily temperature and relative humidity in 2071

### 4.6   Results

By integrating the temperature projections for the next 100 years with the predictive models developed previously, we are now able to offer estimation of the stress exerted on the building's rocks over time. To illustrate this, the Fig. 14 shows the predicted stresses on the rocks in the Courville limestone for each day in 2050.

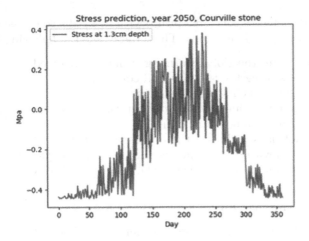

**Fig. 14.** Stress prediction in 2050 for the Courville limestone

## 5   Conclusion

We have shown in this paper that following the behavior of building against weather transformation is a challenging issue and could be well enhanced thanks to the use of machine learning algorithms. Indeed, we have used data analysis algorithms of collected data (during 2 years) from sensors installed on a historical building and we have been able to have precise temperature and humidity values. We have measured the differences between the observations and the data given by official weather centers. These comparisons have been used to adapt the official weather expectations to the building. Finally, we could have better accuracy on the expectation of the building behavior in the future. Then, we have given a framework to predict building transformation for the next years which could help decision-makers to anticipate for urgent works.

## References

1. Doehne, E., Price, C.A.: Stone Conservation: An Overview of Current Research, 2nd edn. Getty Conservation Institute, Canada (2010)
2. Crewdson, M.: Outdoor weathering must verify accelerated testing. Q-Lab Weathering Res. Serv. **1005**, 13 (2008)
3. Lubelli, B., van Hees, R.P., Nijland, T.G.: Salt crystallization damage: how realistic are existing ageing tests?. In: Proceedings of the International Conference on Ageing of Materials and Structures, Delft, pp. 103–111 (2014)
4. McGreevy, J.P.: Thermal properties as controls on rock surface temperature maxima, and possible implications for rock weathering. Earth Surf. Proc. Land. **10**, 125–136 (1985)
5. Warke, P.A., Smith, B.J.: Effects of direct and indirect heating on the validity of rock weathering simulation studies and durability tests. Geomorphology **22**, 347–357 (1998)

6. Smith, B.J., Srinivasan, S., Gomez-Heras, M., Basheer, P.A.M., Viles, H.A.: Near-surface temperature cycling of stone and its implications for scales of surface deterioration. Geomorphology **130**, 76–82 (2011)
7. Huby, E., Thomachot-Schneider, C., Vázquez, P., Fronteau, G., Beck, K.: Experimental thermo-hydric study to simulate natural weathering conditions. J. Cult. Herit. **58**, 12–22 (2022)
8. Weiss, T., Siegesmund, S., Kirchner, D., Sippel, J.: Insolation weathering and hygric dilatation: two competitive factors in stone degradation. Environ. Geol. **46**, 402–413 (2004)
9. Smith, B.J.: Rock temperature measurements from the northwest Sahara and their implications for rock weathering. CATENA **4**, 41–63 (1977)
10. Török, Á.: Oolitic limestone in a polluted atmospheric environment in Budapest: weathering phenomena and alterations in physical properties. Geol. Soc., London, Spec. Publ. **205**(1), 363–379 (2002)
11. Lefèvre, R.A., Ausset, P.: Atmospheric pollution and building materials: stone and glass. Geol. Soc., London, Spec. Publ. **205**(1), 329–345 (2002)
12. Huby, E., Thomachot-Schneider, C., Vázquez, P., Fronteau, G.: Use of microclimatic monitoring to assess potential stone weathering on a monument: example of the Saint-Remi Basilica (Reims, France). Environ. Monit. Assess. **192**, 1–25 (2020)
13. Huby, E.: Réponse de matériaux à des contraintes thermo-hydriques obtenues par suivi climatique: étude de cas de la basilique Saint-Remi de Reims. PhD thesis, URCA-GEGENAA (2021)

# A Novel Image Encryption Technique Using Modified Grain-128

Aissa Belmeguenai[1]($\boxtimes$), Herbadji Djamel[1], Selma Boumerdassi[2], and Berrak Oulaya[3]

[1] Laboratoire de Recherche en Electronique de Skikda,
Université 20 Août 1955- Skikda, BP 26 Route d'El-hadaeik, 21000 Skikda, Algeria
belmeguenaiaissa@yahoo.fr, {a.belmeguenai,d.herbadji}@univ-skikda.dz
[2] Conservatoire National des Arts et Métiers,
92 Rue-Martin, 75141 Paris Cédex 03, France
selma.boumerdassi@inria.fr
[3] Department of Electronics, Faculty of Science and Engineering, Badji Mokhtar
University, LP 12 Annaba, Algeria

**Abstract.** In this work, an alternative version to the stream cipher Grain-128 and an image encryption technique have been proposed. First, the new design is keep the same linear feedback shift register (LFSR) and nonlinear feedback shift register (NLFSR) which are used in Grain-128. The filtering function $h$ of nine variables used in the Grain-128 is replaced by a Boolean function $F$ on fourteen variables. The Boolean function, $F$ is chosen to be balanced, has a correlation immune of order three, has algebraic degree 4 and has nonlinearity 7936. The proposed cipher is implemented to encrypt and decrypt images. The implementation was carried out using Matlab 7.5 platform. The security evaluation of the proposed cipher is presented and compared with others designs. The analysis of the proposed scheme shows superior performance in comparison with previouse designs, and the cipher is secure way for digital data encryption and transmission. The proposed cipher and neural network is combined to build a random number generator for cryptographic applications.

**Keywords:** Boolean function · Grain-128 · Image encryption · Neural network · Stream cipher · Security evaluation

## 1 Introduction

Stream cipher is a secret key encryption system, which combines plain text bits with a pseudo-random bit sequence. Stream ciphers are widely used in many domains (industrial, governmental, telecommunications and individuals), because they are generally fast, and can typically be efficiently implemented in software and much easier in hardware, also it can be implemented much faster than block ciphers and may be much smaller in hardware implementation than block ciphers, they are used in a privileged way in the case of communications

© The Author(s), under exclusive license to Springer Nature Switzerland AG 2024
É. Renault et al. (Eds.): MLN 2023, LNCS 14525, pp. 232–249, 2024.
https://doi.org/10.1007/978-3-031-59933-0_16

likely to be strongly disturbed because they have the advantage of no error propagation, and are particularly suitable for use in environments where no buffering is available and /or plaintext elements need to be processed individually.

The multiplied number of attacks concerning the stream cipher systems based on linear feedback shift registers (combination model or filtering model) [1–9] have led many researchers to be interested to the nonlinear feedback shift registers (NLFSRs); these nonlinear feedback shift registers have very good statistical properties, while remaining concise expressible from a very small amount of bits. They are able to generate sequences of bits of linear complexity which are not linearly bounded by the size of the used register. The NLFSRs also have an interesting additional feature: they allow obtaining a sequence of period $2^L$, where $L$ is the length of register. This allows obtaining a sequence of bits perfectly balanced (which is not the case for the linear feedback shift registers (LFSRs), because the maximum period achieve by LFSR is odd length).

Recent years have witnessed an increase in the research of design and analysis of stream ciphers based on NLFSRs, primarily motivated by eSTREAM, and the ECRYPT stream cipher project [10]. Here we can mention the research works [11–17].

In order to protect against certain attacks as listed in [1–4,7], and [8], we try in this research, to give an alternative version of Grain-128 system for image encryption and decryption. The improved version is based on two binary primitive LFSR, NLFSR and a Boolean combining function. The combining function achieves the best possible trade-offs between algebraic degree, resiliency order and nonlinearity (that is, achieving Siegenthaler's bound and Sarkar et al.'s bound). The proposed design is implemented for image encryption and compared with previouse designs.

The paper is organized as follows. The section two gives a brief description of the Grain-128. In section three we introduce the alternative version of the Grain-128. In section four we present an implantation of the proposed design for digital data encryption. In section five we give the security analysis of the design. Finally section six concludes the paper.

## 2   Grain-128

The Grain-128 [15] is very simple and based on linear feedback shift register (LFSR) of length 128 bits and non linear feedback shift register (NLFSR) of length 128 bits. The content of the LFSR and NLFSR are denoted respectively by $s_{i+1}, s_{i+2}, ..., s_{i+128}$ and $b_{i+1}, b_{i+2}, ..., b_{i+128}$. The output value of the filtering function $h$ of Grain-128 is based on the selected bits from the NLFSR and the LFSR and computed as:

$$h(i) = s_{i+8}b_{i+1} \oplus s_{i+13}s_{i+20} \oplus b_{i+95}s_{i+42} \oplus$$
$$s_{i+60}s_{i+79} \oplus b_{i+12}b_{i+95}s_{i+95}. \tag{1}$$

The Grain-128 produce an output value based of the selected bits of the LFSR and NLFSR states and the output of $h$:

$$Y(i) = \bigoplus_{j \in B} b_{i+j} \oplus h(i) \oplus s_{i+93}. \tag{2}$$

where $B = \{2, 15, 36, 45, 64, 73, 89\}$.

# 3    An Alternative Version of Grain-128

In this section, we give a briefly description of modified Grain-128. The cipher consists of a 128-bit NLFSR, a 128-bit LFSR and a Boolean function $F$. We keep the same LFSR and NLFSR which are used in Grain-128. The filtering function $h$ is replaced by a Boolean function $F$ on 14 variables illustrated by Subsect. 3.1. We denote by $(a_i)_{i \geq 0}$ and $(b_i)_{i \geq 0}$ respectively the output sequences of NLFSR and LFSR. We have the following relation:

$$a_{i+128} = b_i \oplus a_i \oplus a_{i+26} \oplus a_{i+56} \oplus a_{i+91} \oplus$$
$$a_{i+96} \oplus a_{i+3}a_{i+67} \oplus a_{i+11}a_{i+13} \oplus a_{i+17}a_{i+18} \oplus$$
$$a_{i+27}a_{i+59} \oplus a_{i+40}a_{i+48} \oplus a_{i+61}a_{i+65} \oplus a_{i+68}a_{i+84}. \tag{3}$$

$$b_{i+128} = b_i \oplus b_{i+7} \oplus b_{i+38} \oplus b_{i+70} \oplus b_{i+81} \oplus b_{i+96}. \tag{4}$$

The inputs of filtering function $F$ are taken from both LFSR and NLFSR, as:

$$F = a(i+36) \oplus a(i+45) \oplus b(i+95)a(i+64) \oplus$$
$$a(i+36)a(i+64) \oplus f \oplus a(i+12)a(i+95)a(i+64)$$
$$(b(i+95) \oplus a(i+36) \oplus a(i+45)). \tag{5}$$

where

$$f = a(i+2) \oplus a(i+15) \oplus a(i+12)b(i+8) \oplus$$
$$b(i+13)b(i+20) \oplus a(i+95)b(i+42) \oplus$$
$$b(i+60)b(i+79) \oplus a(i+12)a(i+95)b(i+95). \tag{6}$$

The Keystream bits $Z(i)$ is produced by combining the two bits from the state of NLFSR and one bit from the state of LFSR with the output of a non-linear filtering function $F$, as:

$$Z(i) = a(i+73) \oplus a(i+89) \oplus b(i+93) \oplus F(i). \tag{7}$$

## 3.1    Filtering Function

The filtering function $F$ that we used here is drawn from [18]. Let $f = x_1 \oplus x_2 \oplus h$ be a 1-resiliente function on 11 variables has algebraic degree 3 and nonlinearity 960. Where $h = x_3x_4 \oplus x_5x_6 \oplus x_7x_8 \oplus x_9x_{10} \oplus x_3x_7x_{11}$ is the Boolean function proposed for Grain-128 [14] has nonlinearity 240. Let $f^*$ be a Boolean function generated from $f$ by replacing the variable $x_{11}$ with $(x_{12} \oplus x_{13})$. Let $f_1 = x_{12} \oplus x_{13} \oplus f$ and $f_2 = x_{11} \oplus x_{13} \oplus f^*$ be two Boolean functions on $F_2^{13}$. We construct the function $F$ in 14-variable in the following way: $F = (1 \oplus x_{14}) f_1 \oplus x_{14} f_2 = f_1 \oplus x_{14}(f_1 \oplus f_2) = f_1 \oplus x_{14}(x_{11} \oplus x_{12} \oplus f \oplus f^*) = x_{12} \oplus x_{13} \oplus f \oplus x_{11} x_{14} \oplus x_{12} x_{14} \oplus x_3 x_7 x_{14}(x_{11} \oplus x_{12} \oplus x_{13})$. Then the function $F$ is balanced, has a correlation immune of order three, has algebraic degree 4 and has nonlinearity 7936.

**Table 1.** Comparison Between Modified Grain-128 and Grain-128

| Cryptographic Criteria | Grain-128 | Modified Grain-128 |
|---|---|---|
| Number of Variables | 17 | 17 |
| Balancedness | Balanced | Balanced |
| Algebraic Degree | 3 | 4 |
| Correlation Immune | 7 | 6 |
| Nonlinearity | 61440 | 63488 |
| Linear Complexity | at least $\binom{128}{3}$ | at least $\binom{128}{4}$ |
| Gate AND | 6 | 11 |
| Gate XOR | 12 | 16 |

# 4   Implementation

Recently, many encryption technique based on chaotic theory have been proposed. In this section an implementation of the modified Grain-128 for image encryption was introduced. The implementation was carried out using Matlab 7.5 platform.

By simulating how the human brain works, neural network algorithms aim to identify causal correlations in a batch of data [19]. This study generates secret key sequences by using the newly built modified Grain-128 as its transformation function. With a single input layer and output layer, the INN consists of three hidden layers. The input layer gets input from the previous output iteratively. The starting points of the trajectory are distributed randomly around the space using the transformation functions. The output layer and hidden layer each use the LFSR and NLFSR once. Additionally, the suggested approach produces 128 pseudorandom sequence values in a single cycle see Fig. 1.

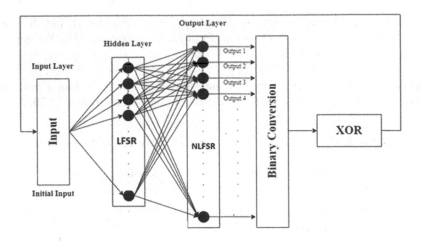

**Fig. 1.** Proposed customized neural network for random number generation.

## 4.1   Encryption Algorithm

To encrypt the original images $I_{L \times C}$, we generate a pseudo-random matrix $Z_N$. The modified Grain-128, is used to generate pseudo-random matrix, by iterating $N$ times, using the Eq. 7. Where $N = \frac{L}{2} \times \frac{C}{2} \times 8$, $L$ and $C$ the row and the column of original image $I_{L \times C}$. The block diagram of the proposed encryption methode is illustrate by Figs. 2 and 3. The encryption direction of each element of the image as follows:

**Step 1**: Divide the original image $I_{L \times C}$ into four equal elements $I_{1,N}$, $I_{2,N}$, $I_{3,N}$ and $I_{4,N}$.

**Step 2**: Encrypting each element with the element which is adjoined to it with $\oplus$ operation to obtain cipher image elements $I_{1,N}^*$, $I_{2,N}^*$, $I_{3,N}^*$ and $I_{4,N}^*$. The ciphering direction of each element is defined from the other element as shown in Fig. 2. To encrypt the content of image elements, we calculate:

$I_{1,N}^* = I_{1,N} \oplus I_{2,N}$
$I_{2,N}^* = I_{2,N} \oplus I_{3,N}$
$I_{3,N}^* = I_{3,N} \oplus I_{4,N}$
$I_{4,N}^* = I_{1,N} \oplus I_{4,N}.$

**Step 3**: Encryption of the content of each element of the image resulting from the previous step with $\oplus$ operation to obtain cipher image elements $q_{1,N}$, $q_{2,N}$, $q_{3,N}$ and $q_{4,N}$. The ciphering direction of each element is defined from the other part as shown in Figs. 2 as:

$$q_{1,N} = I^*_{1,N} \oplus Z_N$$
$$q_{2,N} = I^*_{2,N} \oplus q_{1,N}$$
$$q_{3,N} = I^*_{3,N} \oplus q_{2,N}$$
$$q_{4,N} = I^*_{4,N} \oplus q_{3,N}.$$

Where $\oplus$ is the XOR operation between two element of the image, and $q_{1,N}$, $q_{2,N}$, $q_{3,N}$, $q_{4,N}$ are the encrypted image elements.

**Step 4**: Combining encrypted image elements.

## 4.2 Decryption Algorithm

**Step 1**: Divide the encrypted image $C_{L \times C}$ into four equal elements $q_{1,N}$, $q_{2,N}$, $q_{3,N}$ and $q_{4,N}$.

**Step 2**: Decryption of the content of each element of the encrypted image by the following relations (stage 1 of decrypted): $I^*_{4,N} = q_{4,N} \oplus I_{3,N}$

$$I^*_{3,N} = q_{3,N} \oplus q_{2,N}$$
$$I^*_{2,N} = q_{2,N} \oplus q_{1,N}$$
$$I^*_{1,N} = q_{1,N} \oplus Z_N.$$

**Step 3**: stage 2 of decrypted

$$I_{4,N} = I^*_{4,N} \oplus I_{1,N}$$
$$I_{3,N} = I^*_{3,N} \oplus I_{4,N}$$
$$I_{2,N} = I^*_{2,N} \oplus I_{3,N}$$
$$I_{1,N} = I^*_{1,N} \oplus I_{2,N}.$$

**Step 4**: Combining decrypted image elements

## 4.3 Visual Test

Visual test and histograms of original images and encrypted images are compared in Figs. 4, 5, 6 and 7, there is no visual information observed in the encrypted images. It is clear that the histograms of the encrypted images are fairly uniform and significantly different from the histograms of the original images, this is implies that the statistic attacks will be inefficient.

# 5  Security Evaluation of Modified Grain-128

The correlation coefficient, information entropy, Mean Absolute Error (MAE), number of pixels change rate (NPCR), unified average changing intensity (UACI) and Peak signal to noise ratio (PSNR) are used for the evaluation of improved Grain-128. Refer to [20]. In order to evaluate the modified Grain-128, a number of evaluation parameters like, correlation coefficient, information entropy, NPCR, UACI, MAE and PSNR, are taken for validate the modified Grain-128. The proposed scheme is compared to a Grain-128 [15, 21–25] and [26].

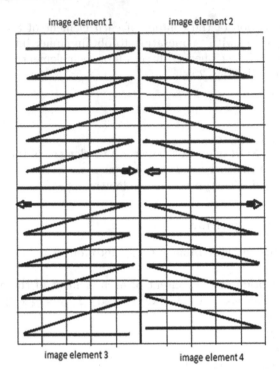

**Fig. 2.** Direction of diffusing each pixels image elements.

## 5.1   Noise Analysis

When acquiring or transmitting the image a noise is added to the image. A good cryptosystem should be resist to the noise attack.Suppose the encrypted image is impaired by a noise as: $I_B^* = I^* + B$ where $I_B^*$ and $I^*$ are the noisy encrypted image and the encrypted image, respectively. $B$ is a noise generated with function $randn$ in Matlab. It is a matrix of the same size as the encrypted image containing pseudo-random values drawn from a normal distribution with zero-mean and identity standard deviation. The results of noise analysis are shown in Fig. 12. The noise has no influence on the decryption process.

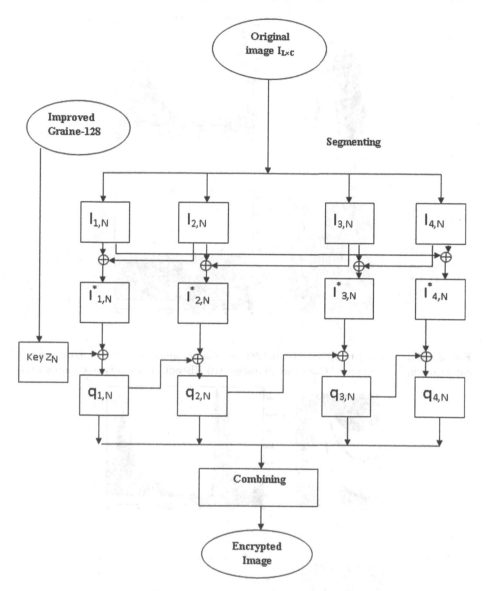

**Fig. 3.** Block diagram of proposed image encryption scheme.

## 5.2 Correlation Coefficient

We present the correlation coefficient analysis on improved Grain-128. The correlation coefficient between two horizontally adjacent pixels, two vertically adjacent pixels and two diagonally adjacent pixels in original and encrypted images were tested. Tests were performed on Cameraman and Banana images. Figures 8 and 9 show respectively the correlation distribution of two horizontally adjacent

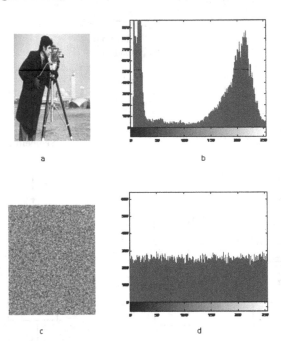

**Fig. 4.** Test 1 using modified Grain-128: (a) Cameraman image, (b) Histogram of cameraman, (c) Encrypted Cameraman image, (d) Histogram of encrypted cameraman.

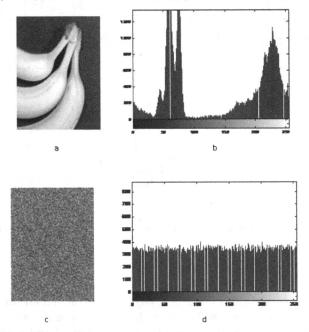

**Fig. 5.** Test 2 using modified Grain-128: (a) Banana image, (b) Histogram of Banana, (c) Encrypted Banana image, (d) Histogram of encrypted Banana.

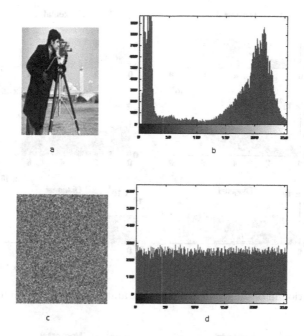

**Fig. 6.** Test 3 using Grain-128: (a) Cameraman image, (b) Histogram of cameraman, (c) Encrypted Cameraman image, (d) Histogram of encrypted cameraman.

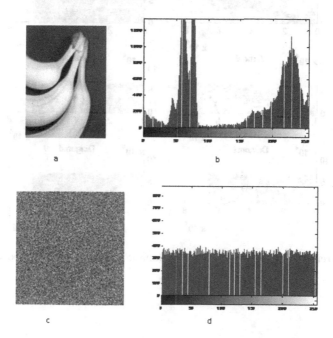

**Fig. 7.** Test 4 using Grain-128: (a) Banana image, (b) Histogram of Banana, (c) Encrypted Banana image, (d) Histogram of encrypted Banana.

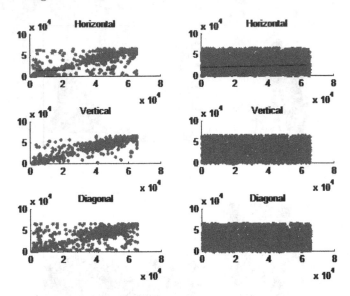

**Fig. 8.** Correlation in original and cipher Cameraman image using modified Grain-128.

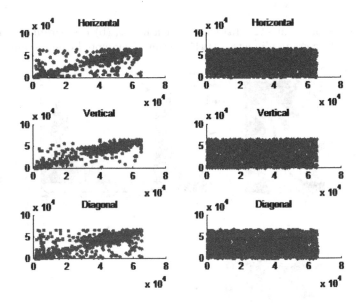

**Fig. 9.** Correlation in original and cipher Banana image using modified Grain-128.

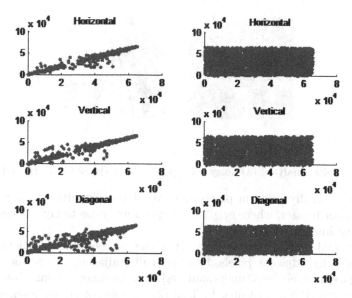

**Fig. 10.** Correlation in original and cipher Cameraman image using Grain-128.

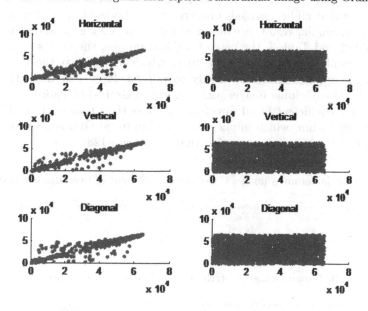

**Fig. 11.** Correlation in original and cipher Banana image using Grain-128.

a                                      b

**Fig. 12.** Noise Analysis: (a) noisy encrypted image, (b) noisy decrypted image.

pixels, two vertically adjacent pixels and two diagonally adjacent pixels in original and cipher images, where original images were respectively the Cameraman and Banana images.

Table 2 and Table 3 respectively give the correlation coefficient between two horizontally adjacent pixels, two vertically adjacent pixels and two diagonally adjacent pixels for Cameraman and Banana images using Grain-128 and improved Grain-128 algorithms. In Table 2, the case of cipher images, the values of horizontal, vertical and diagonal correlation coefficient are near to zero, which means that cipher image is uncorrelated in horizontal, vertical and diagonal direction. Similar results are obtained for Banana image as shown in Table 3. From Table 2 and Table 3, the case of original images, the values of correlation coefficient in all directions are close to 1, which means that the two adjacent pixels in the original image are highly correlated. It is clear from Table 2 and Table 3, the case of cipher images using improved Grain-128 algorithm, the value of correlation coefficient in all directions are less than those obtained by using Grain-128 algorithm, which implies that the improved Grain-128 is an efficient and secure way for image encryption than the Grain-128.

**Table 2.** Cameraman image: Correlation coefficient of two adjacent pixels.

| Direction of adjacent pixels | Original image [15] | Original image proposed | Cipher image [15] | Cipher image proposed | Cipher image [22] |
|---|---|---|---|---|---|
| Horizontal | 0.9319 | 0.9319 | −0.0027 | 0.00057047 | 0.0047 |
| Vertical | 0.9661 | 0.9661 | 0.0077 | −0.00089904 | −0.0066 |
| Diagonal | 0.9104 | 0.9104 | 0.0014 | −0.0013 | 0.0031 |

**Table 3.** Banana image: Correlation coefficient of two adjacent pixels.

| Direction of adjacent pixels | Original image [15] | Original image proposed scheme | Cipher image [15] | Cipher image proposed scheem |
|---|---|---|---|---|
| Horizontal | 0.9928 | 0.9928 | 0.000146 | −0.00010889 |
| Vertical | 0.9915 | 0.9915 | 0.0030 | 0.0013 |
| Diagonal | 0.9798 | 0.9798 | −0.00093 | 0.00081312 |

## 5.3   Information Entropy

In this section, information entropy analysis, on Grain-128, improved Grain-128, Achterbahn-128 [21] and improved Achterbahn-128 [22] is explored. Simulation results for entropy analysis are shown in Table 4. For all algorithms, the value of entropy is very close to theoretical value of 8 bits. But Grain-128, Achterbahn-128 [21] and improved Achterbahn-128 [22] have less entropy as compare to improved Grain-128. This means that information leakage is negligible and improved Grain-128 encryption algorithm is secure against entropy attack than the Grain-128, Achterbahn-128 and improved Achterbahn-128.

**Table 4.** Entropy

| Cipher image | [15] | [21] | [22] | proposed scheme | [23] | [26] | [24] |
|---|---|---|---|---|---|---|---|
| Cameraman | 7.9971 | 7.9972 | 7.9974 | 7.9975 | 7.9971 | 7.9972 | 7.9904 |
| Banana | 7.9981 | 7.9979 | 7.9981 | 7.9982 | - | - | - |

## 5.4 NPCR , UACI and MAE

NPCR , UACI and MAE are three common measures used to check the influence of a one pixel change on the overall image to resist differential attack. The NPCR , UACI and MAE are calculated by using Grain-128, improved Grain-128, Achterbahn-128 and improved Achterbahn-128 algorithms. Simulation results are shown in Tables 5, 6 and 7. The NPCR , UACI and MAE values are higher, the encryption algorithm is better.

From Tables 5, 6 and 7, it is clear that the improved Grain-128 has good diffusion characteristics than Graine-128, Achterbahn-128 and improved Achterbahn-128. The results in Tables 5, 6 and 7, show that the improved Grain-128 has strong diffusion mechanism as compare with the Grain-128, Achterbahn-128 and the improved Achterbahn-128.

**Table 5.** NPCR (%) results

| Image | [15] | [21] | [22] | Proposed | [23] | [24] | [26] |
|---|---|---|---|---|---|---|---|
| Cameraman | 99.56 | 99.60 | 99.58 | 99.59 | 99.55 | 99.60 | 99.67 |
| Banana | 99.57 | 99.57 | 99.59 | 99.60 | - | - | - |

**Table 6.** UACI (%) results

| Image | [15] | [21] | [22] | Proposed | [23] | [24] | [26] |
|---|---|---|---|---|---|---|---|
| Cameraman | 36.96 | 36.94 | 37.08 | 37.08 | 33.44 | 33.15 | 33.66 |
| Banana | 35.40 | 35.39 | 35.41 | 35.44 | - | - | - |

**Table 7.** MAE results

| Cipher image | [15] | [21] | [22] | Proposed |
|---|---|---|---|---|
| Cameraman | 40.28 | 40.409 | 40.41 | 40.63 |
| Banana | 39.26 | 39.36 | 39.40 | 39.48 |

## 5.5    PSNR (dB)

Peak signal to noise ratio (PSNR) is measure for evaluate an encryption scheme. PSNR reflects the encryption quality. The PSNR value is lower, the encryption algorithm is better. Simulation results of (PSNR) using Graine-128, improved Grain-128, Achterbahn-128 and improved Achterbahn-128 algorithms are shown in Table 8. From Table 8, it is clear that the values obtained by improved Grain-128 are less than those obtained by Grain-128, Achterbahn-128 and improved Achterbahn-128.

**Table 8.** PSNR (dB) results

| Image | [15] | [21] | [22] | Proposed | [25] | [26] |
|---|---|---|---|---|---|---|
| Cameraman | 6.9635 | 6.953 | 6.9492 | 6.94 | 8.2978 | 7.3786 |
| Banana | 7.31 | 7.31 | 7.3 | 7.28 | - | - |

## 5.6    Time-Memory Trade-Off Attack

A generic time-memory-data trade-off attack (TMTOA) on stream ciphers costs $O(2^{k/2})$, where $k$ is the size of the key, [16]. The proposed cipher has the secret key 256-bit. Thus, the expected complexity of a time-memory-data trade-off attack is not lower than $O(2^{128})$.

## 5.7    Berlekamps-Massey Algorithm

The Berlekamps-Massey algorithm [2] requires $2 \times \binom{128}{4}$ successive data to work. In order to prevent a Berlekamp-Massey algorithm, the keystream generator must produce a pseudo-random sequence with linear complexity highest possible. The proposed keystream generator produce a pseudo-random sequence with linear complexity at least equal to $\binom{128}{4}$, this means that it is completely excluding to use the Berlekamp-Massey algorithm.

## 5.8   Algebraic Attack

In the case of a stream cipher based on the combination generator, the combiner function must be balanced, possess high algebraic degree, high order of correlation immunity, high nonlinearity and high algebraic immunity to resist a several attacks [1–9].

In the case of a filter generator, a high order of correlation immunity is not necessary, the correlation immunity of order 1 is considered sufficient [27]. It is important to increase the algebraic degree and nonlinearity to resist to the attacks [28].

The functions $F$ is of algebraic degree 4 has order of correlation 3 and nonlinearity 7936, it takes seven bits from NLFSR and seven bits from LFSR. The output of NLFSR to introduce nonlinearity together with the $F$, explain by the term $a(i+12)a(i+95)a(i+64)(b(i+95) \oplus a(i+36) \oplus a(i+45))$ in the Algebraic normal form (ANF) of $F$. Solving an equations for the initial 256 bits state is not possible due to the nonlinear update of the NLFSR, which make the algebraic degree of the output of $F$ large in general and also varying in time. This will make difficult any algebraic attack on the cipher. In order to increase the nonlinearity of the keystream sequence $Z(i)$. The output of $F$ is combined linearly with two bits of NLFSR and a bit of LFSR. So the output $Z(i)$ has in total nonlinearity $2^3 \times 7936 = 63488$ this is greater than of the nonlinearity of Grain-128. Table 1 shows the comparison between cryptography properties of the improved Grain-128 and of Grain-128.

## 6   Conclusion

In this work, an alternative design to the Grain-128 and an image encryption technique have been proposed. The proposed technique has high resistance against the potential attacks. The performance of the proposed work is analyzed by comparing it with with the Grain-128, Achterbahn-128 and the improved Achterbahn-128 algorithms and it is found that the proposed work can provide more secure information, and has a good diffusion characteristics and is secure way for image encryption and transmission. In future work, The proposed cipher and neural network are combined to build a random number generator for cryptographic applications.

## References

1. Massey, J.L.: Shift-Register synthesis and BCH decoding. IEEE Trans. Inf. Theory **IT-15**, 122–127 (1969)
2. Berlekamp, R.: Algebraic Coding Theory. Grow- Hill, New- York (1968)
3. Siegenthaler, T.: Decrypting a class of stream ciphers using cipher text only. IEEE Trans. Comput. C-34, N°1, 81–85 (1985)
4. Meier W., Staffelbach, O.: Fast correlation attacks on stream chiper. In : Advances in Cryptology- EUROCRYPT 1988, éd. Par GÜNTHER (C.G), Lectures Notes in Computer science N° 430, pp. 301–314, Springer, Cham (1988)

5. Zeng, K., Hung, M.: On the linear syndrome method in cryptanalysis. In: Gold-wasser, S. (ed.) CRYPTO 1988. LNCS, vol. 403, pp. 469–478. Springer, New York (1990). https://doi.org/10.1007/0-387-34799-2_32

6. Zeng, K., Yang, C.H., Rao, T.R.N.: An improved linear syndrome algorithm in cryptanalysis with applications. In: Menezes, A.J., Vanstone, S.A. (eds.) CRYPTO 1990. LNCS, vol. 537, pp. 34–47. Springer, Heidelberg (1991). https://doi.org/10.1007/3-540-38424-3_3

7. Golić, J.D.: Linear cryptanalysis of stream ciphers. In: Preneel, B. (ed.) FSE 1994. LNCS, vol. 1008, pp. 154–169. Springer, Heidelberg (1995). https://doi.org/10.1007/3-540-60590-8_13

8. Courtois, N.T., Meier, W.: Algebraic attacks on stream ciphers with linear feed-back. In: Biham, E. (ed.) EUROCRYPT 2003. LNCS, vol. 2656, pp. 345–359. Springer, Heidelberg (2003). https://doi.org/10.1007/3-540-39200-9_21

9. Courtois, N.T.: Fast algebraic attacks on stream ciphers with linear feedback. In: Boneh, D. (ed.) CRYPTO 2003. LNCS, vol. 2729, pp. 176–194. Springer, Heidelberg (2003). https://doi.org/10.1007/978-3-540-45146-4_11

10. eSTREAM, the ECRYPT Stream Cipher Project. http://www.ecrypt.eu.org/stream/

11. Gammel, B.M., Gottfert, R., Kniffler, O.: The Achterbahn stream cipher, eSTREAM, ECRYPT Stream Cipher Project, Report 2005/002, 29 April 2005. https://www.ecrypt.eu.org/stream/papers.html

12. Gammel, B.M., Gottfert, R., Kniffler, O.: Status of Achterbahn and Tweaks. In: SASC 2006—Stream Ciphers Revisited (Leuven, Belgium, 2-3 February 2006), Workshop Record, pp. 302–315 (2006)

13. Gammel, B.M., Gottfert, R., Kniffler, O.: Achterbahn-128/80, eSTREAM, ECRYPT Stream Cipher Project, Report 2006/001 (2006)

14. Hell, M., Johansson, T., Meier, W.: Grain a stream cipher for constrained environments. Int. J. Wirel. Mob. Comput. Spec. Issue Secur. Comput. Netw. Mob. Syst. (2006)

15. Hell, M., Johansson, T., Meier, W.: A stream cipher proposal: Grain-128. In: IEEE International Symposium on Information Theory (ISIT 2006) (2006)

16. De Cannière, C., Preneel, B.: Trivium a stream cipher construction inspired by block cipher design principles. In: eSTREAM,ECRYPT Stream Cipher Project, Report 2005/030 (2005-04-29) (2005)

17. Cid, C., Kiyomoto, S., Kurihara, J.: The RAKAPOSHI stream cipher. In: Qing, S., Mitchell, C.J., Wang, G. (eds.) ICICS 2009. LNCS, vol. 5927, pp. 32–46. Springer, Heidelberg (2009). https://doi.org/10.1007/978-3-642-11145-7_5

18. Belmeguenai, A., Ouchtati, S., Zennir, Y.: An alternative method of construction of resilient functions. Int. J. Comput. Commun. 12, 7–9 (2018)

19. Patel, S., et al.: Colour image encryption based on customized neural network and DNA encoding. Neural Comput. Appl. 33, 14533–14550 (2021). https://doi.org/10.1007/s00521-021-06096-2

20. Ahmad, J., Ahmed, F.: Efficiency analysis and security evaluation of image encryption schemes. Proc. Int. J. Video Image Process. Netw. Secur. IJVIPNS-IJENS 12(04)

21. Belmeguenai, A., Berrak, O., Mansouri, K.: Security evaluation and implementation of Achterbahn-128 for images encryption. In: Advances in Computer Science, Proceedings of the 6th European Conference of Computer Science (ECCS 2015), pp. 232–238 (2015)

22. Belmeguenai, A., Berrak, O., Mansouri, K.: Image encryption using improved keystream generator of achterbahn-128. In: Proceedings of the 11th Joint Conference on Computer Vision, Imaging and Computer Graphics Theory and Applications (VISIGRAPP 2016). Volume 3: VISAPP, pp. 333–339 (2016)

23. Wang, X., Zhu, X., Zhang, Y.: An image encryption algorithm based on Josephus traversing and mixed chaotic map. IEEE Access **6**, 23733–23746 (2018)

24. Mozaffari, S.: Parallel image encryption with bitplane decomposi- tion and genetic algorithm. Multimedia Tools Appl. **77**(19), 25799–25819 (2018)

25. Khalid, I., Jamal, S.S., Shah, T., Shah, D., Hazzazi, M.M.: A novel scheme of image encryption based on elliptic curves isomorphism and substitution boxes. IEEE Access **9**, 77798–77810 (2021)

26. Ramzan, M., Shah, T., Hazzzi, M.M., Aljedi, A., Alharbi, A.R.: Construction of s-boxes using different maps over elliptic curves for image encryption. IEEE Access (2021)

27. Carlet, C., Feng, K.: An infinite class of balanced functions with optimal algebraic immunity, good immunity to fast algebraic attacks and good nonlinearity. In: Pieprzyk, J. (ed.) ASIACRYPT 2008. LNCS, vol. 5350, pp. 425–440. Springer, Heidelberg (2008). https://doi.org/10.1007/978-3-540-89255-7_26

28. Canteaut, A., Trabbia, M.: Improved fast correlation attacks using parity-check equations of weight 4 and 5. In: Preneel, B. (ed.) EUROCRYPT 2000. LNCS, vol. 1807, pp. 573–588. Springer, Heidelberg (2000). https://doi.org/10.1007/3-540-45539-6_40

# Transformation Network Model for Ear Recognition

Aimee Booysens and Serestina Viriri[✉]

School of Mathematics, Statistics and Computer Science,
University of KwaZulu-Natal, Durban, South Africa
210501411@stu.ukzn.ac.za, viriris@ukzn.ac.za

**Abstract.** Biometrics is the recognition of a human using biometric characteristics for identification, which may be physiological or behavioural. The physiological biometric features are the face, ear, iris, fingerprint and handprint; behavioural biometrics are signatures, voice, gait pattern and keystrokes. Numerous systems have been developed to distinguish biometric traits used in multiple applications, such as forensic investigations and security systems. With the current worldwide pandemic, facial identification has failed due to users wearing masks; however, the human ear has proven more suitable as it is visible. This paper presents the main contribution to presenting the results of a CNN developed using Transfer Learning to pre-train the CNN before applying a Transformer Network. The performance achieved in this research shows the efficiency of the Transformer Network on ear recognition. The experiments showed that Transformer Network achieved the best accuracy of 92.60% and 92.56% with epochs of 50 and 90.

**Keywords:** Ear Biometrics · Ear Recognition · Transformer Network · Machine Learning

## 1 Introduction

The ear begins to develop in a fetus during the fifth and seventh weeks of pregnancy [2]. At this stage, the face acquires a more distinguishable shape as the mouth, nostrils, and ears begin to form. There is still no exact timeline at which the outer ear is created, but it is accepted that a cluster of embryonic cells connects to establish the ear. These are called auricular hillocks, which begin growing in the lower portion of the neck. The auricular hillocks broaden and inter and twine within the seventh week to deliver the ear's shape. Within the ninth week, the hillocks move to the ear canal and are more noticeable as the ear [2].

Biometrics is the recognition of a human using their biometric characteristics, which may be physiological or behavioural. The physiological biometric features are the DNA, face, ear, facial, iris, fingerprint, hand geometry, hand vein and palm print, and behavioural biometrics are signatures, and gait path-connected

E. Renault et al. (Eds.): MLN 2023, LNCS 14525, pp. 250–266, 2024.
https://doi.org/10.1007/978-3-031-59933-0_17

keystrokes. Voice is considered a combination of biometric and physiological characteristics. Numerous systems have been developed to distinguish biometric traits, which have been used in multiple applications, such as forensic investigations and security systems. With the current Worldwide pandemic, facial identification has failed due to users wearing masks. However, the human ear has proven more suitable as it is visible.

In different physiological biometric qualities, the ear has received much consideration of late as it tends to be said that it is a solid biometric for human acknowledgement [6] Ear biometric framework is dependable as it does not change, is of uniform tone, and its position is fixed at the centre of the face's side. The size of an individual's ear is more critical than a unique finger impression and makes it simpler to capture an image of the subject without necessarily needing to gain information from the subject, [6]. There are numerous difficulties in correctly gauging the details of the ear, these are concealment of the ear by clothing, hair, ear ornaments and jewellery. Another interference could be the different angle that the image was taken, concealing essential characteristics of the ear's anatomy. These difficulties have made ear recognition a secondary role in identification systems and techniques commonly used for identification and verification.

Although several computer-aided detection models have been developed to identify ears, low accuracy and sensitivity are still significant concerns that misidentify ears. Existing models are also computationally complex and expensive. In this paper, an ear recognition model based on Transformer Network is proposed.

The remaining work is structured as follows: Sect. 2 presents related works, and Sect. 3 presents detailed data and methodology explored in this study. The experimental results and discussion are provided in Sect. 4, and Sect. 5 concludes the paper.

## 2    Related Work

This section presents different algorithms using the Convolutional Neural Network (CNN) for ear identifications, a summary of the related works is shown in Table 1.

The competition Emeršič et al. [9] organised the dataset of the UERC, which was used for the bench-mark, training and testing sets. In the completion, it was seen that handcrafted feature extraction methods, such as LBP [29] and patterns of oriented edge magnitudes (POEM) [28], and CNN-based feature extraction methods were used to obtain the ear identification. The challenges were to find methods to remove occlusions such as earrings, hair, other obstacles, and background from the ear image. The occlusion was done by creating a binary ear mask, and then the system recognition was done using the handcrafted features. Another proposed approach was to calculate the score of matrices from the CNN-based features and handcrafted features when they are fused, a 30% detection rate was achieved.

Tian et al. [26] applied a CNN to ear recognition in which they designed a CNN - it was made up of three convolutional layers, a fully connected layer, and a softmax classifier. The database used was USTB ear, which consisted of 79 subjects with various pose angles. The images utilised excluded earrings, headsets, or similar occlusions. Chowdhury et al. [8] proposed an ear biometric recognition system that uses local features of the ear and then uses a neural network to identify the ear. The method estimates where the ear could be in the input image and then takes the edge features from the identified ear. After identifying the ear, a neural network matches the extracted feature with a feature database. The databases used in this system were AMI, WPUT, IITD, and UERC, which achieved an accuracy of 70.58%, 67.01%, 81.98%, and 57.75%.

Raveane, Galdámez and Arrieta [24] presented that it is difficult to precisely detect and locate an ear within an image, this challenge increases when working with the variable condition and this could also be because of the odd shape of the human ears as well as lighting conditions and the changing profile shape of an ear when photographed, [24]. The ear detection system used multiple CNN's, combined with a detection grouping algorithm, to identify an ear's presence and location. The proposed method matches other methods' performance when analysed against clean and purpose-shot photographs, reaching an accuracy of upwards of 98%. It outperforms them with a rate of over 86% when the system is subjected to non-cooperative natural images where the subject appears in challenging orientations and photographic conditions.

Multiple Scale Faster Region-based Convolutional Neural Networks (Faster R-CNN) to detect ears from 2D profile images was proposed by Zhang and Mu [32]. This method was used by taking three regions of different scales that are detected to defer the information from the ear location within the context of the ear in the image, which was done to extract the ear correctly. The system was tested with 200 web images that achieved a 98% accuracy. Other experiments conducted were on the Collection J2 of the University of Notre Dame Biometrics Database (UND-J2) and University of Beira Interior Ear dataset (UBEAR); these achieved a detection rate of 100% and 98.22%, respectively, but these datasets contained large occlusions, scale and pose variation.

Kohlakala and Coetzer [18] presented semi-automated and fully automated ear-based biometric verification systems. CNN and morphological postprocessing manually identify the ear region. It is used to classify ears in the image's foreground or background. The binary contour image applied the matching for feature extraction, and this was done by implementing a Euclidean distance measure, which had a ranking to verify for authentication. The Mathematical Analysis of Images ear database and the Indian Institute of Technology Delhi ear database were two databases, which achieved 99.20% and 96.06%, respectively.

Geometric deep learning (GDL) generalises CNNs to non-Euclidean domains, presented by [27] Tomczyk and Szczepaniak. It used convolutional filters with a mixture of Gaussian models. These filters were used so that the images could be easily rotated without interpolation. It shows the published experimental results that the approach did the rotational equivalence property to detect rotated structures. Still, it does not need labour-intensive training on all rotated and non-rotated images.

Hammam et al. [5] presented and compared ear recognition models built with handcrafted and CNN features. The paper took seven, performing handcrafted descriptors to extract the discriminating ear image. They then took the extracted ear and trained it using Support Vector Machines (SVM) to learn a suitable model. They then used CNN-based models, which used a variant of the AlexNet architecture. The results obtained on three ear datasets showed the CNN-based models' performance increased by 22%. This paper also investigated if the left and right ears have symmetry. The results obtained by the two datasets indicate a high impact of balance between the ears.

Alkababji and Mohammed [4] presented the use of a Deep Learning item detector called faster region-based convolutional neural networks (Faster R-CNN) for ear detection. This CNN is used for feature extraction. It used the Principal Component Analysis (PCA) and a genetic algorithm for feature reduction and selection. It also used a connected artificial neural network as the matcher. The results achieved an accuracy of 97.8% success rate.

Jamil et al. [17] build and train a CNN model for ear biometrics in various uniform illuminations measured using lumens. They considered that their work was the first to test the performance of CNN on underexposed or overexposed images. The results showed that images with uniform illumination with a luminance of above 25 lux, achieved a result of 100%. The CNN model had problems recognising images when the lux was below ten, but produced an accuracy of 97%. This result shows that CNN architecture performs just as well as the other systems. It was found that the dataset had rotations which affected the results.

Hansley et al. [15] presented an unconstrained ear recognition framework that was better than the current state-of-the-art systems using publicly available databases. They developed CNN-based solutions for ear normalisation and description. This was done using a handcrafted descriptor. The published experimental results show This was done in two stages. The first stage was to find the landmark detectors, which were untrained scenarios. The next step was to generate a geometric image normalisation to boost the performance. It was seen that the CNN descriptor was better than other CNN-based works in the literature. The obtained results were higher than different reported results for the UERC challenge.

**Table 1.** Summary of the related works

| Author | Dataset | Accuracy | Summary |
|---|---|---|---|
| Zhang and Mu [32] | Notre Dame Biometrics Database and University of Beira Interior Ear dataset | 100 & 98.22 | This system contained large occlusions, scale and pose variation. |
| Kohlakala and Coetzer [18] | Mathematical Analysis of Images ear database and Indian Institute of Technology Delhi ear database | 99.2 & 96.06 | It is used to classify ears either in the foreground or background of the image. The binary contour image applied the matching for feature extraction, and this was done by implementing a Euclidean distance measure, which had a ranking to verify for authentication. |
| Tomczyk and Szczepaniak [27] | NA | NA | It shows the published experimental results that the approach did the rotation equivalence property to detect rotated structures. |
| Hammam et al. [5] | Three ear datasets but not stated | 22 | The paper took seven performing handcrafted descriptors to extract the discriminating ear image. They then took the extracted ear and trained it using Support Vector Machines (SVM) to learn a suitable model. |
| Alkababji and Mohammed [4] | NA | 97.8 | It used the Principal Component Analysis (PCA) and a genetic algorithm for feature reduction and selection. |
| Jamil et al. [17] | Very underexposed or overexposed database | 97 | They considered that their work was the first to test the performance of CNN on very underexposed or overexposed images. |
| Hansley et al. [15] | UERC challenge | NA | This was done using handcrafted descriptors, which were fused to improve recognition. |
| Tian et al. [26] | AMI, WPUT, IITD, and UERC | 70.58, 67.01, 81.98, & 57.75 | This system used deep convolutional neural network (CNN) to ear recognition. There were occlusions like no earrings, headsets, or similar occlusions. |
| Raveane, Galdámez and Arrieta [24] | NA | 98 | This system used variable conditions and this could also be because of the odd shape of the human ears and changing lighting conditions. |
| Emeršič et al. [9] | NA | 30 | It was a handcrafted feature extraction methods, such as LBP and patterns of oriented edge magnitudes (POEM), and CNN-based feature extraction methods were used to obtain the ear identification. |

# 3    Data and Methods

## 3.1    Dataset

In this study, all the experiments were performed with numerous public ear datasets an explanation of these datasets is provided below. UBEAR, EarVN1.0, IIT, ITWE and AWE databases are best suited for ear identification due to their large data size. However, it shows that EarVN1.0 has the foremost prominent usage during age estimation using CNN techniques. It is an appropriate dataset for ear images taken in a controlled environment, while ITWE is compatible with classifying ears in an uncontrolled environment, a summary of the datasets is shown in Table 3.

**Mathematical Analysis of Images (AMI) Ear Database.** The AMI ear database [14] was collected at the University of Las Palmas. The database comprises 700 ear images of 100 distinct Caucasian male and female adults between the ages of 19 and 65. All images within the database were taken under an equivalent illumination and a glued camera position. Both the left- and right-hand sides of the ears were captured. The pictures obtained were cropped to form the ear area covering almost half the image. The pose of the images varies in yaw and servery in pitch angles, and this dataset is often found publicly.

**The Indian Institute of Technology (IIT) Delhi Ear Database.** The IIT database [19] was collected by the Indian Institute of Technology Delhi in New Delhi between October 2006 and June 2007. The database is formed from 421 images of 121 distinct adults of both male and female. All images were taken inside the environment, with no significant occlusions present, and only the right-hand side of the ear was captured. The pictures obtained in the dataset were both raw and normalised. The normalised images were in greyscale and of size $272 \times 204$ pixels.

**The University of Beira Ear Database (UBEAR).** The University of Beira presented the UBEAR database [23]. The database comprises 4429 images of 126 subjects, and these were of both males and females. The images were taken under varying lighting conditions, angles and partial occlusions were present. These images were of both the left- and right-hand side of the ear.

**The Annotated Web Ear Database (AWE).** The AWE ear database [11] is a set of public figures from web images. The database was formed from 1000 images of 100 different subjects whose sizes vary and were tightly cropped. Both the left- and right-hand sides of the ears were taken.

**EarVN1.0.** The EarVN1.0 database [16] comprises, 28412 images of 164 Asian male and female subjects, left- and right-hand sides of the ear were captured. Collection was during 2018 and is formed from unconstrained conditions, including camera systems and lighting conditions. The pictures are cropped from facial images to obtain the ears, and the pictures have significant variations in pose, scale and illumination.

**The Western Pomeranian University of Technology Ear Database (WPUTE).** The Western Pomeranian University of Technology Ear Database WPUTE [13] was obtained in the year 2010 to gauge the ear recognition performance for images obtained within the wild. The database contains 2071 ear images belonging to 501 subjects. The images were of various sizes and were of both the left- and right-hand sides of the ear, these were taken under different indoor lighting conditions and rotations. There were some occlusions included in the database, these were the headset, earrings and hearing aids.

**The Unconstrained Ear Recognition Challenge (UERC).** The Unconstrained Ear Recognition Challenge (UERC) database [10] was obtained in 2017, then extended in 2019 and is a mix of two databases that currently exist and a newly created one. The database contains 3706 subjects with, 11804 ear images, and the database ears have both right- and left-hand side images.

**In the Wild Ear Database (ITWE).** The In the Wild Ear Database (ITWE) [12] was created for recognition evaluation and has, 2058 total images, 231 male and female subjects. A boundary box obtained these images of the ear, and the coordinates of those boundary boxes were released with the gathering. The pictures contained cluttering backgrounds and were of variable size and determination. The database includes both left- and right-hand sides of the ear, but no differentiation was given about the ears.

**The University of Science and Technology, Beijing (USTB) Ear Database.** The University of Science and Technology Beijing (USTB) Ear Database [31] contained cropped ear and head profile images of male and female subjects split into four sets. Dataset one includes 60 subjects and has 180 images of right close-up ears during 2002. These images were taken under different lighting and experienced some shearing and rotation. Data set two contains 77 subjects and has 308 images of the right-hand side ear approximately 2 m away from the ear and were taken in 2004. These images were taken under different lighting conditions. Dataset three contains 103 subjects and has 1600 images, these images were taken during the year 2004. The images are on the proper and left rotation, and therefore the images are of the dimensions $768 \times 576$ pixels. The dataset contains, 25500 images of 500 subjects; these were obtained from 2007 to 2008; the subject was in the centre of the camera circle. The images were taken when the subject looked upwards, downwards and at eye level. The images during this dataset contained different yaw and pitch poses. The databases are available on request and accessible for research.

**The Carreira-Perpinan (CP) Ear Database.** The Carreira-Perpinan (CP) [7] ear database is an early dataset of the ear utilised for ear recognition systems. It was created in 1995 and contained 102 images with 17 subjects. The images were captured in a controlled environment, and therefore the images include variability in minor pose variation.

**The Indian Institute of Technology Kanpur (IITK) Ear Database.** The Indian Institute of Technology Kanpur (IITK) is an ear database [22] that the Institute of Technology of Kanpur compiled. The database is split into three sets, the first set consists of 190 male and female subjects of profile images. The total number of images was 801. The second dataset also contained 801, with a total of 89 subjects, these images had variations in pitch angle. The third dataset contains 1070 images of an equivalent of 89 subjects, but with a variation in yaw and angle.

**The Forensic Ear Identification Database (FEARID).** The Forensic Ear Identification Database (FEARID) database [3] is different from other databases as it only includes ear prints. These contain no occlusions, variable angles, or illumination. Though there is no mention of any variables, other influences like the force the ear was pressed against the scanner and the scanner's cleanliness need to be considered. This database comprised, 7364 images of 1229 subjects. This database was used for forensic applications and not for biometric use.

**The University of Notre Dame (UND) Database.** The University of Notre Dame (UND) database contains [30] many subsets of 2D and 3D ear images. These images were appropriated over a period from 2003 to 2005. The database contains, 3480 3D images from 952 male and female subjects and 464, 2D images from 114 male and female subjects. These images were taken in different lighting conditions, yaw, pitch poses and angles. The images are only of the left-hand side of the ear.

**The Face Recognition Technology Database (FERET).** The Face Recognition Technology Database (FERET) [21] is a sizeable facial image database, and was obtained between the years 1995 to 1996. It contains 1564 subjects and has a total of 14126 images. These images were collected for face recognition and were of the left- and right-hand profile images, which made them perfect for 2D ear recognition.

**The Pose, Illumination and Expression (PIE).** Carnegie Mellon University obtained The Pose, Illumination and Expression database [25], which contains,

**Table 2.** Summary of Datasets

| | Database | Year | Number of subjects | Number of Images | Left Ear Count | Right Ear Count | Total Ears | Image Size | Country | Sides |
|---|---|---|---|---|---|---|---|---|---|---|
| 1 | Institute of Technology Delhi Ear Database (IIT Delhi-I) [19] | 2007 | 121 | 471 | | 471 | 471 | 272 × 204 | India | Right |
| | Institute of Technology Delhi Ear Database (IIT Delhi-II) [19] | NA | 221 | 793 | | 793 | 793 | 272 × 204 | India | Right |
| 2 | The University of Science & Technology Beijing (USTB Ear I) [31] | 2002 | 60 | 185 | | 185 | 185 | Varied | China | Right |
| | The University of Science & Technology Beijing (USTB Ear II) [31] | 2004 | 77 | 308 | | 308 | 308 | Varied | China | Right |
| 3 | The Annotated Web Ears database (AWE) [11] | 2016 | 100 | 1000 | 500 | 500 | 1000 | Varied | Slovenia | Both |
| | The Annotated Web Ears database extended (AWE extend) [11] | 2017 | 346 | 4104 | 2052 | 2052 | 4104 | Varied | Slovenia | Both |
| 4 | Mathematical Analysis of Images Ear Database (AMI) [14] | NA | 106 | 700 | 420 | 280 | 700 | 492 × 702 | Spain | Both |
| 5 | The West Pomeranian University of Technology Ear Database (WPUTE) [13] | 2010 | 501 | 2071 | 829 | 1242 | 2071 | Varied | Poland | Both |
| 6 | Unconstrained Ear Recognition Challenge database (UERC) [10] | 2017 | 3706 | 11804 | 5902 | 5902 | 11804 | Varied | Solvenia | Both |
| 7 | EarVN1.0 [16] | 2018 | 164 | 28412 | 14206 | 14206 | 28412 | Varied and low resolution | Vietnam | Both |
| 8 | The In-the wild Ear Database (TWE) [12] | 2015 | 55 | 605 | 424 | 181 | 605 | Varied | Solvenia | Both |
| 9 | The Carreira-Perpinan (CP) [7] | 1995 | 17 | 102 | 102 | | 102 | Varied | NA | Left |
| 10 | The University of Beira Ear Database (UBEAR) [23] | 2011 | 126 | 4430 | 2215 | 2215 | 4430 | 1280 × 960 | Mozambique | Both |
| 11 | Indian Institute of Technology Kanpur (IITK) [22] | 2011 | 801 | 190 | 95 | 95 | 190 | Varied | India | Both |
| 12 | The Forensic Ear Identification Database (FEARID) [3] | 2005 | 1229 | 1229 | 615 | 614 | 1229 | Varied | United Kingdom, Italy and Netherlands. | Both |
| 13 | University of Notre Dame (UND) [30] | 2006 | 3480 | 952 | 952 | | 952 | Varied | France | Left |

**Table 3.** Summary of Datasets

| | Database | Year | Number of subjects | Number of Images | Left Ear Count | Right Ear Count | Total Ears | Image Size | Country | Sides |
|---|---|---|---|---|---|---|---|---|---|---|
| 14 | The Face Recognition Technology Database (FERET) [21] | 2010 | 9427 | 4745 | 3796 | 949 | 4745 | Varied | Spain | Both |
| 15 | The Pose, Illumination and Expression (PIE) [25] | 2002 | 40000 | 68 | 34 | 34 | 68 | Varied | USA | Both |
| 16 | The XM2VTS Ear Database [20] | NA | 2360 | 295 | 89 | 206 | 295 | 720 × 576 | UK | Both |
| 17 | The West Virginia University (WVU) [1] | 2006 | 460 | 402 | 402 | | 402 | Varied | USA | Left |

40000 images and 68 subjects. The images are of the facial profile and have different poses, illuminations and expressions.

**The XM2VTS Ear Database.** The XM2VTS ear database [20] is frontal and profiles facial images from the University of Surrey; the database contains 295 subjects and, 2360 images captured during controlled conditions. These images were a set of cropped images 720 × 576 pixel size and were from video data.

**The West Virginia University (WVU) Ear Database.** The West Virginia University (WVU) Ear Database [1] is a video database and is formed from 137 subjects. The system was an advanced capturing procedure that allowed them to capture the ear at different angles; these images included earrings and eyeglasses.

### 3.2  Pre-processing

Image pre-processing is a considerable part of the deep-learning task. Most CNN models generally require a large dataset to learn to discriminate features suitably for making predictions and obtaining a good performance. As images in the datasets are of different sizes, the inputted images need to be resized to conform to all the other CNN models, but the features need to be preserved when resizing is performed.

### 3.3  Transfer Learning

In this study, the concept of transfer learning was adopted and helped with the pre-trained CNN model for large datasets to learn features of the target (right and left ears). It will transfer the features learned by the deep CNN models on other CNN models to this dataset. The number of deep CNN model parameters increases as the network gets deeper, which is used to achieve improved efficiency.

Hence, it requires many datasets for training, making it computationally complex and applying these models directly on small and new dataset results in feature extraction bias, overfitting, and poor generalisation. The pre-trained

CNN modified and fine-tuned its structure to suit the dataset given. This concept of transfer learning is computationally expensive, has less training time, overcomes limitations of the dataset, improves performance, and is faster than training a model from the beginning. The pretraining CNN model fine-tuned in this work is the Transformer Network. The proposed structure is represented in Fig. 1.

### 3.4    Transformer Network Architecture

The Transformer Network is an encoder-decoder architecture based on attention layers. The difference between a Convolutional Neural Network and the Transformer Network is that the data can be passed in parallel, this means that the GPU can be utilised effectively and efficiently. The speed of the training is also increased by processing it in parallel. It is seen that the Transformation Network is based on a multi-headed attention layer and by doing this the vanishing gradient issue is overcome.

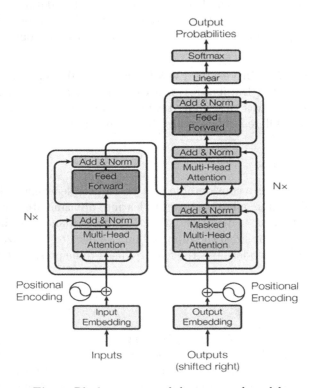

**Fig. 1.** Block structure of the proposed model

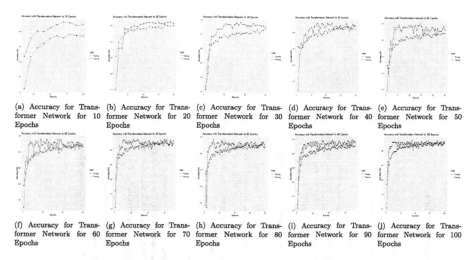

(a) Accuracy for Transformer Network for 10 Epochs

(b) Accuracy for Transformer Network for 20 Epochs

(c) Accuracy for Transformer Network for 30 Epochs

(d) Accuracy for Transformer Network for 40 Epochs

(e) Accuracy for Transformer Network for 50 Epochs

(f) Accuracy for Transformer Network for 60 Epochs

(g) Accuracy for Transformer Network for 70 Epochs

(h) Accuracy for Transformer Network for 80 Epochs

(i) Accuracy for Transformer Network for 90 Epochs

(j) Accuracy for Transformer Network for 100 Epochs

**Fig. 2.** Accuracy for the ear dataset of each Transformer Network

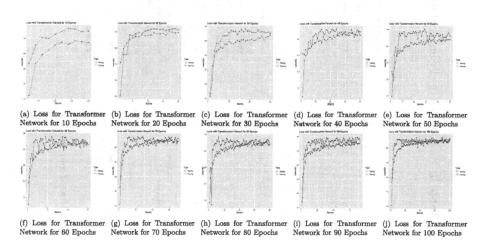

(a) Loss for Transformer Network for 10 Epochs

(b) Loss for Transformer Network for 20 Epochs

(c) Loss for Transformer Network for 30 Epochs

(d) Loss for Transformer Network for 40 Epochs

(e) Loss for Transformer Network for 50 Epochs

(f) Loss for Transformer Network for 60 Epochs

(g) Loss for Transformer Network for 70 Epochs

(h) Loss for Transformer Network for 80 Epochs

(i) Loss for Transformer Network for 90 Epochs

(j) Loss for Transformer Network for 100 Epochs

**Fig. 3.** Loss for the ear dataset of each Transformer Network

**Table 4.** Proposed method compared with the related studies

| Author | Dataset | Accuracy | Summary |
|---|---|---|---|
| Emeršič et al. [9] | NA | 30 | It used handcrafted feature extraction methods, such as LBP, POEM and CNN-based feature extraction methods were used to obtain the ear identification. |
| Tian et al. [26] | AMI, WPUT, IITD, and UERC | 70.58, 67.01, 81.98, & 57.75 | This system used deep CNN to do ear recognition. There were occlusions like no earrings, headsets, or similar occlusions. |
| Raveane, Galdámez and Arrieta [24] | NA | 98 | This system used variable conditions due to the odd shape human ear and changing lighting conditions. |
| Zhang and Mu [32] | UND and UBEAR | 100 & 98.22 | This system contained large occlusions, scale and pose variation. |
| Kohlakala and Coetzer [18] | AMI and IIT-Delhi | 99.2 & 96.06 | It is used to classify ears either in the foreground or background of the image. The binary contour image applied the matching for feature extraction, and this was done by implementing a Euclidean distance measure, which had a ranking to verify for authentication. |
| Tomczyk and Szczepaniak [27] | NA | NA | It shows the published experimental results that the approach did the rotation equivalence property to detect rotated structures. |
| Hammam et al. [5] | Three ear datasets but not stated | 22 | The paper took seven, performing handcrafted descriptors to extract the discriminating ear image. Then took the extracted ear and trained it using SVM to learn a suitable model. |
| Alkababji and Mohammed [4] | NA | 97.8 | It used the PCA and a genetic algorithm for feature reduction and selection. |
| Jamil et al. [17] | Very underexposed or overexposed database | 97 | This work was the first to test the performance of CNN on very underexposed or overexposed images. |
| Hansley et al. [15] | UERC challenge | NA | This was done using handcrafted descriptors, which were fused to improve recognition. |
| **Our Work** | **AWE, AMI and IIT** | **92** | |

## 4    Results and Discussion

Transformer Network variants were fine-tuned on all the ear datasets to detect the ear. Each dataset is split into 20% training and 80% test sets. The experiments were entirely performed using Keras deep learning framework using the TensorFlow backend. The models were evaluated using the popular evaluation metrics, Eq. 1 -5, (accuracy, recall and precision. The performances of all experiments are evaluated by using a series of confusion matrix-based performance metrics.

The confusion matrices used to evaluate the classifiers, with true positives (TP) representing the ears that are correctly classified as positive, true negatives (TN) representing the ears that are correctly classified as negative, false positives (FP) representing the ears that are incorrectly classified as positive, and false negatives (FN) representing the ears being incorrectly classified as negative.

**Precision.** It is the ratio of correctly classified negative instances by a model to the overall number of true negative instances being tested, Eq. 3.

**Accuracy.** It is a measure that indicates the ratio of all the correctly recognized cases to the overall number of cases. While this metric generally gives a decent reflection of the classifier, it may not reflect a classifier's true performance in a swhichrio where there is an uneven class distribution. Accuracy can be computed uslng the following formula, Eq. 1.

**Recall.** It is the ratio of all correctly classified positive instances by a model to the overall number of positive classifications by a model. A low precision indicates that a model suffers from high false positives. Precision can be computed using the following formula, Eq. 2.

$$Accuracy = \frac{TP + TN}{TP + FP + TN + FN} \tag{1}$$

$$Recall = \frac{TP}{TP + FN} \tag{2}$$

$$Precision = \frac{TN}{TN + FP} \tag{3}$$

$$TPR = Recall = \frac{TP}{TP + FN} \tag{4}$$

$$FPR = 1 - Precision = \frac{FP}{FP + TN} \tag{5}$$

The results obtained are presented in Figs. 2 and 3 this is the accuracy and loss of these datasets. In the Transformer Network at different epochs the accuracy is determined using the test set. The models performed at extracting and learning discriminative features from the dataset. Transformer Network with 50 and 90 Epochs attains the best accuracy 92.60 and 92.56%, and the Transformer Network results are noted in Table 5.

An advantage of Transformer Networks is that they are smaller with fewer parameters, faster, and obtain transfer learning successfully from the datasets. The worst performing was 20 epochs, as shown in Table 5. The reason that this performed poorly could have been because it did not have enough data to learn from. This was done to conform to the model's image input size. It can be seen that performance improves as the model gets deeper. On average it was seen that overfitting occurred at 30 iterations and stabilised at around 50. The best performing Global Transformer Network is at epochs 50 and 90, as shown in Table 5 and, this is because of the large number of parameters. It began to converge from the 30 iterations and then stabilised until 50 iterations when overfitting started. Determining the most suitable hyperparameters was one of the challenges faced as the overfitting, which was limited due to the data samples. The results of the proposed methods compared with related studies are presented in Table 4.

**Table 5.** Performance of Transformer Network

| Epochs | Accuracy (%) | Loss (%) |
|--------|--------------|----------|
| 10     | 90.42        | 47.78    |
| 20     | 87.06        | 37.47    |
| 30     | 91.10        | 31.30    |
| 40     | 90.97        | 30.94    |
| 50     | 92.60        | 27.96    |
| 60     | 91.74        | 28.03    |
| 70     | 91.81        | 26.80    |
| 80     | 92.18        | 26.67    |
| 90     | 92.56        | 25.42    |
| 100    | 91.91        | 26.91    |

## 5   Conclusion

This study investigated and implemented did pre-process by fine-tuning and pre-training the CNN before applying the Transformer Network to automatically identify ears on the most prominent and publicly available datasets. Transformer Networks that achieved state-of-the-art performance over other architectures to maximise accuracy and efficiency were explored and fine-tuned on profile images. The fine-tuning technique is valuable to utilise rich generic features learned from significant datasets sources such as ImageNet to complement the lack of annotated datasets affecting ear domains. The experimental results show the effectiveness of Transformer Network in extracting and learning distinctive features from the ear images and then classifying them into a left or right-suitable class. Out of the ten Transformer Network variants explored in this study, the Transformer Network with 90 Epochs outperformed the others, as evident in Table 4. One of the limitations found was that it is easily over-fitted. To overcome this, you need to have compelling image preprocessing techniques. Although the proposed methodology is specified to do ear detection, it could be extended to detect other parts of the face, given the right set of datasets.

# References

1. Abaza, A.: High Performance Image Processing Techniques in Automated Identification Systems. West Virginia University, Morgantown (2008)
2. Abaza, A., Ross, A., Hebert, C., Harrison, M.A.F., Nixon, M.S.: A survey on ear biometrics. ACM Comput. Surv. **45**(2) (2013). https://doi.org/10.1145/2431211.2431221
3. Alberink, I., Ruifrok, A.: Performance of the fearid earprint identification system. Forensic Sci. Int. **166**(2–3), 145–154 (2007)
4. Alkababji, A.M., Mohammed, O.H.: Real time ear recognition using deep learning. Telkomnika **19**(2), 523–530 (2021)
5. Alshazly, H., Linse, C., Barth, E., Martinetz, T.: Handcrafted versus CNN features for ear recognition. Symmetry **11**(12), 1493 (2019)
6. Chen, H., Bhanu, B.: Ear Biometrics 3D, pp. 241–248. Springer, US, Boston, MA (2009). https://doi.org/10.1145/2431211.2431221
7. Carreira-Perpinan, M.A.: Compression neural networks for feature extraction: application to human recognition from ear images (1995)
8. Chowdhury, D.P., Bakshi, S., Guo, G., Sa, P.K.: On applicability of tunable filter bank based feature for ear biometrics: a study from constrained to unconstrained. J. Med. Syst. **42**(1), 1–20 (2018)
9. Emeršič, Ž., Štepec, D., Štruc, V., Peer, P.: Training convolutional neural networks with limited training data for ear recognition in the wild. arXiv preprint arXiv:1711.09952 (2017)
10. Emeršič, Ž., et al.: The unconstrained ear recognition challenge. In: 2017 IEEE International Joint Conference on Biometrics (IJCB), pp. 715–724. IEEE (2017)
11. Emeršič, Ž, Štruc, V., Peer, P.: Ear recognition: more than a survey. Neurocomputing **255**, 26–39 (2017)
12. Emeršič, V., Peer, P.: Ear biometric database in the wild, pp. 27–32 (2015). https://doi.org/10.1109/IWOBI.2015.7160139
13. Frejlichowski, D., Tyszkiewicz, N.: The west Pomeranian university of technology ear database - a tool for testing biometric algorithms
14. Gonzalez, E., Alvarez, L., Mazorra, L.: Ami ear database (2012). http://ctim.ulpgc.es/research_works/ami_ear_database/
15. Hansley, E.E., Segundo, M.P., Sarkar, S.: Employing fusion of learned and handcrafted features for unconstrained ear recognition. IET Biometrics **7**(3), 215–223 (2018)
16. Hoang, V.T.: Earvn1.0: a new large-scale ear images dataset in the wild. Data Brief **27**, 104630 (2019).https://doi.org/10.1016/j.dib.2019.104630, https://www.sciencedirect.com/science/article/pii/S2352340919309850

17. Jamil, N., Almisreb, A., Ariffin, S.M.Z.S.Z., Din, N.M., Hamzah, R.: Can convolution neural network (CNN) triumph in ear recognition of uniform illumination invariant? (2018)
18. Kohlakala, A., Coetzer, J.: Ear-based biometric authentication through the detection of prominent contours. SAIEE Africa Res. J. **112**(2), 89–98 (2021)
19. Kumar, A.: Iit delhi ear database version 1.0. (2007). http://webold.iitd.ac.in/~biometrics/Database_Ear.htm
20. Messer, K., et al.: XM2VTSDB: the extended M2VTS database. In: Second International Conference on Audio and Video-based Biometric Person Authentication, vol. 964, pp. 965–966. Citeseer (1999)
21. Phillips, P., Wechsler, H., Huang, J., Rauss, P.J.: The FERET database and evaluation procedure for face-recognition algorithms. Image Vis. Comput. **16**(5), 295–306 (1998)
22. Prakash, S., Jayaraman, U., Gupta, P.: Connected component based technique for automatic ear detection. In: 2009 16th IEEE International Conference on Image Processing (ICIP), pp. 2741–2744. IEEE (2009)
23. Raposo, R., Hoyle, E., Peixinho, A., ProenÄsa, H.: Ubear: A dataset of ear images captured on-the-move in uncontrolled conditions (2011).https://doi.org/10.1109/CIBIM.2011.5949208
24. Raveane, W., Galdamez, P.L., Gonzalez Arrieta, M.A.: Ear detection and localization with convolutional neural networks in natural images and videos. Processes **7**(7), 457 (2019)
25. Sim, T., Baker, S., Bsat, M.: The CMU pose, illumination, and expression (pie) database of human faces. Technical report CMU-RI-TR-01-02, Carnegie Mellon University, Pittsburgh, PA (2001)
26. Tian, L., Mu, Z.: Ear recognition based on deep convolutional network. In: 2016 9th International Congress on Image and Signal Processing, BioMedical Engineering and Informatics (CISP-BMEI), pp. 437–441. IEEE (2016)
27. Tomczyk, A., Szczepaniak, P.S.: Ear detection using convolutional neural network on graphs with filter rotation. Sensors **19**(24), 5510 (2019)
28. Vu, N.S., Dee, H.M., Caplier, A.: Face recognition using the poem descriptor. Pattern Recogn. **45**(7), 2478–2488 (2012)
29. Wang, Z.Q., Yan, X.D.: Multi-scale feature extraction algorithm of ear image. In: 2011 International Conference on Electric Information and Control Engineering, pp. 528–531. IEEE (2011)
30. Yan, P., Bowyer, K.: Empirical evaluation of advanced ear biometrics. In: 2005 IEEE Computer Society Conference on Computer Vision and Pattern Recognition (CVPR 2005)-Workshops, pp. 41–41. IEEE (2005)
31. Zhang, Y., Mu, Z.C., Yuan, L., Yu, C., Qing, L.: USTB-Helloear: a large database of Ear Images Photographed Under Uncontrolled Conditions, pp. 405–416 (2017). https://doi.org/10.1007/978-3-319-71589-6_35
32. Zhang, Y., Mu, Z.: Ear detection under uncontrolled conditions with multiple scale faster region-based convolutional neural networks. Symmetry **9**(4), 53 (2017)

# Cybersecurity Analytics: Toward an Efficient ML-Based Network Intrusion Detection System (NIDS)

Tariq Mouatassim[1]([✉]), Hassan El Ghazi[1], Khadija Bouzaachane[2], El Mahdi El Guarmah[2,3], and Iyad Lahsen-Cherif[1]

[1] AGNOX Lab, INPT, Rabat, Morocco
Tariqmouatassim@gmail.com
[2] L2IS, Cadi Ayyad University, FST, Marrakesh, Morocco
[3] Mathematics and Informatics Department, Royal Moroccan Air Force Academy, Marrakesh, Morocco

**Abstract.** ML-based NIDS are among the tools used within the framework of Cybersecurity analytics to tackle intrusions and alert for potential or ongoing cyberattacks. Their design relies heavily on precollected datasets on which ML algorithms are trained. However, NIDS datasets are often confronted with two major problems: imbalanced classes and outdated traffic flows. In fact, in one hand, designing NIDSs using obsolete datasets (like KDD99 and NSLKDD) may result in poor performances when implemented in nowadays network environment, due to their lack of modern attack styles and recent normal traffic scenarios. On the other hand, a high imbalance ratio could result in decrease of the efficiency of NIDS, especially for rarely encountered attack types. Therefore, in this study, binary and multiclass intrusion detection models are proposed, using Tree-based algorithms: Decision Tree(DT), Random Forest(RF), ExtraTrees, Gradient Boosting(GB), Adaboost, and XGBoost algorithms. The main advantage of this work is the use of a recent and well-ranked dataset, NF-UQ-NIDS-v2, which have been balanced using k-means undersampling, to train Tree-based models for intrusion detection. Through the experiments, we found that our approach presented satisfactory prediction time and performances, with low rates of false negatives and false positives.

**Keywords:** Cybersecurity analytics · Network Intrusion Detection Systems · NF-UQ-NIDS-v2 · K-means clustering · Tree-based algorithms · hyper-parameter Bayesian Optimization Tree Parzen Estimator

## 1 Introduction

In recent years, due to the extensive use of Information Technology (IT) and Internet, the number of networked computers has been increasing in our daily lives. This new reality has brought risks, which threaten availability, confidentiality as well as integrity of valuable information in case of a cyberattack. To

É. Renault et al. (Eds.): MLN 2023, LNCS 14525, pp. 267–284, 2024.
https://doi.org/10.1007/978-3-031-59933-0_18

protect against it, organisation are adopting cybersecurity analytics, a new holistic approach to deal with an ever-growing amount of threats. This approach relies on Machine Learning (ML) techniques to handle vast quantity of data with the aim of ensuring a full protection of IT systems and networks. Among solutions which are designed by training ML algorithms on a precollected dataset, NIDS stands as one of the most preferred protection mechanisms. In order to improve efficiency of a NIDS in the conception stage, several parameters could be taking into consideration. The choice of a recent and high quality dataset for training the models combined to a good preprocessing of the dataset are among the most important success keys in building machine learning models for intrusion detection. In this paper, we propose a new approach of preprocessing, based on proportionally sampling the dataset using K-means clustering. The sampled dataset was trained using six different Tree-based machine-learning algorithms. Finally, the results obtained in the present work were optimized using BO-TPE.

The rest of this paper is organized as follows. In the next section, Cybersecurity analytics approach is detailed with a special attention to ML tools. Section 3 outlines literature review on the most interesting works dealing with ML-based NIDS published between 2018 and 2023 . Our proposed solution is presented in Sect. 4. The experimental results are discussed in Sect. 5 and the optimization methods and results are presented in Sect. 6. Finally, conclusions are drawn, and directions for future works are suggested in Sect. 7.

## 2    Cybersecurity Analytics and ML Tools

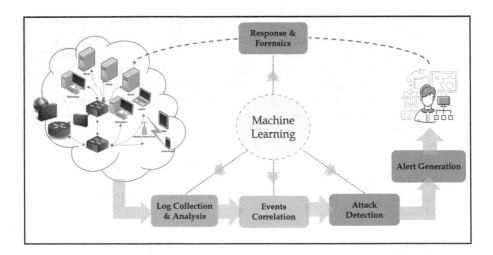

**Fig. 1.** Cybersecurity analytics process

Cybersecurity analytics is a proactive approach, wherein no longer organisations wait until cyberattacks happen, but rather prepare to face cyberthreats by monitoring and correlating in real-time events which could potential involve malicious activities. The emphasis is put, firstly, in collecting logs from IT equipment. Then logs are forwarded to security systems which are responsible for predicting, detecting and identifying cyberattacks. Hence, whenever a threat is predicted or a cyberattack took place, incident response procedures are performed accordingly to contain and eradicate eventual damages, then restore normal operations. Finally, Digital Forensics are conducted to extract digital evidences admissible in court to incriminate the ones responsible for the cyberattacks.

Throughout the cybersecurity analytics process, Fig. 1, vast quantity of data is generated and need to be processed adequately to enable a full protection of the system. Accordingly, ML and data analytics are leveraged to develop security solutions, such as AI powered Security Information and Event Management systems (SIEM). These systems are comprehensive tools designed to prevent, detect, and react against cyberattacks [10]. They could make decisions and take action instantaneously based on events occurring within the supervised network. Therefore, this enhances detection rates and minimizes false positives rates while reducing manual intervention [13].

To develop ML based security solutions, researchers follow the process provided in Fig. 2. First, the dataset is preprocessed (cleansing, feature selection, normalisation,etc.), then split into a training set and a testing set. Next, the training set is used to train a ML algorithm so as to obtain a model which is evaluated on the testing set. Once validated, the model could be applied to real data to give a prediction or to classify input data into categories.

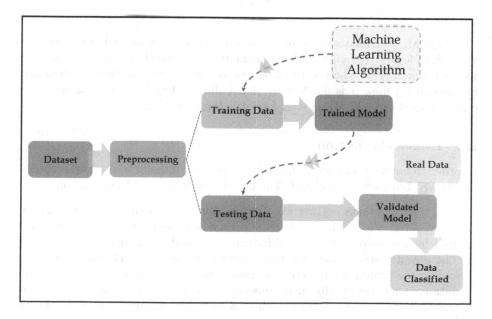

**Fig. 2.** ML process

## 2.1    Log Collection and Analysis

Prior to detecting cyberattacks, logs from various sources ( firewalls, network equipment, servers, endpoints, etc.) should be collected to a central server. Usually, logs are overwhelming semi-structured text including a timestamp and a description of an event, such as a successive logging failure of a user from an IP address to a system. Yet, manually analysing and parsing logs to obtain structured data, using regular expressions as well as search engines, is impracticable. Example of a parsed network log is given in table 1

**Table 1.** Example of a parsed network log

| IPV4_SRC_ADDR | 192.168.10.35 |
|---|---|
| L4_SRC_PORT | 1305 |
| IPV4_DST_ADDR | 172.16.18.200 |
| L4_DST_PORT | 21 |
| IN_BYTES | 9 |
| OUT_BYTES | 129 |
| TCP_FLAGS | 24 |
| CLIENT_TCP FLAGS | 24 |
| MAX_TTL | 32 |
| ICMP_IPV_TYPE | 0 |

Thus, ML techniques come to the rescue to offer automated collection and analysis of logs. An example of ML-driven anomaly detection based on log analysis is given in [14] , where authors proposed a framework to identify anomalies in massive log files. It is noteworthy that collected logs are the main source of datasets used to build ML models for intrusion detection.

## 2.2    Events Correlation

In order to improve security monitoring, logs from different sources aren't taken separately, but rather correlated. The correlation logic could depends on [11]:

- The time of events occurrence. For instance, a scanning activity bringing about a NIDS log alert, followed by a successive logging failure event generated by a web application could indicate an eventual intrusion attempt.
- The physical area. As an example, a user physically present in a data-center and in the same time remotely accessing to a server belonging to the same data-center is potentially suspicious events. The first log event could be generated by an access control system or a surveillance camera, and the second log event is generated by the computer.
- The user involved. For example, a user simultaneously accessing physically and remotely a device could reveal a malevolent action.

Correlation can be performed visually by using different windows and different graphical solutions, or automated using either if/else rules or ML techniques [11].

## 2.3   Attack Detection

One of the main purposes of Cybersecurity analytics is to detect cyberattacks promptly. To achieve this objective, organisations took advantages of ML and data analytics to get ready for attack scenarios. Accordingly, malware detection as well as insider threat and intrusion detection mechanisms are developed. The intrusion detection is the process of monitoring the events occurring in a computer system or network and analyzing them for signs of possible incidents, which are violations or imminent threats of violation of computer security policies, acceptable use policies, or standard security practices [28].

Intrusion detection systems could be either implemented on a host or in the network. In the first case, Host-based Intrusion Detection System (HIDS) monitors host system for potential malicious activity. To this end, the executed processes, inbound and outbound connections as well as changes in the registry, are continuously analyzed for any sign of anomaly. In the second case, NIDS monitors packets on a network and analyzes traffic looking for potential intrusions. They are generally located in transit areas, where the greatest amount of network traffic flows, and make decisions based, either on a specific signature spotted in the traffic or on an anomaly detected in the flow [8]. Although anomaly-based NIDS can detect efficiently unknown network attacks, there still a big challenge to ensure high accurate performances by maximizing the detection rate, accuracy and precision [9].

## 2.4   Incident Response and Forensics

The aim of incident response framework is to ensure that appropriate actions are taken in order to minimize loss or theft of information [3]. Incident response lifecycle involve preparation, detection and analysis, containment, eradication and recovery, and post-incident activities. when incident response team are facing major cyberattacks, they have to proceed to incident triage to identify and prioritize responses.

Again, ML have been leveraged in this field. In [22], authors proposed a new approach for a quick incident triage. After applying dimensionality reduction to enable intelligible visualisation of events, k-means clustering were chosen for triage and trained on unlabeled data.

After incidents is closed and lessons are learned, forensics teams intervene to perform digital investigation. When forensics procedures are carried out by human experts, they often require a great deal of time, are applied to massive quantity of forensic data, and are prone to errors [7]. For this reason, ML tools are used to boost investigative precision and efficiency. For example, k-means clustering may be utilized to detect anomalies in extracted evidences (logs, volatile memories, drive images) [23].

**Table 2.** Performances and limitations of related works

| Reference | Year | Dataset | Algorithms | Accuracy/FAR/FPR | Limitations |
|---|---|---|---|---|---|
| [17] | 2018 | UNSW-NB15 | NB ; DT ; ARM ; ANN | Accuracy: between 63.97% and 93.23% FPR: between 6.77% and 36.03% | Imbalanced dataset |
| [20] | 2019 | UNSW-NB15 | LR ; GB ; SVM | Accuracy: 86.04% for multiclass classification | Imbalanced dataset |
| [26] | 2019 | KDD-99 | LR ; GB ; RF ; XGBoost | AUC of 96,62% Training time 24 s for XGBoost | Dataset obsolete |
| [15] | 2020 | CSE-CIC-IDS2018 | KNN ; RF ; GB ; Adaboost ; DT ; LDA | Accuracy of Adaboost 99.69% | several classes had low accuracy performances |
| [4] | 2020 | NSL-KDD | XGBoost-DNN ; LR ; NB ; SVM | maximum Accuracy XGBoost-DNN 97% | outdated attack scenarios in the dataset |
| [16] | 2020 | UNSW-NB15 | LR ; KNN ; ANN ; DT ; SVM | Accuracy: between 53.43% and 82.66% multiclass classification | imbalanced dataset |
| [18] | 2021 | UNSW-NB15 | DT(C5, CHAID, CART, QUEST) | Accuracy: 88.92% classification | just 4 out of 9 types of attacks detected, imbalanced dataset |
| [1] | 2022 | TON_IoT ; posemat IoT-23, IoT-ID | RF ; LR ; DT ; GNB | Accuracy for binary classification 99,62% | |
| [29] | 2023 | KDDCUP'99 ; CIC-MalMem-2022 | XGBoost-DNN ; LR ; NB ; SVM | accuracy of 99.99% (KDDCUP'99) and 100% (CIC-MalMem-2022) | shortcomings in used datasets (KDDCUP'99 is outdated, CIC-MalMem-2022 is specific for malware detection) |

## 3   Related Works

In this section, relevant papers related to ML-driven anomaly-based NIDS are presented.

Some authors trained their models on relatively recent but imbalanced dataset, like UNSW-NB15. In fact, part of this later was used by Koroniotis et al. (2018) [17] and helped develop a framework for the detection of botnet attacks using ML algorithms, namely Association Rule Mining (ARM), Artificial Neural Network (ANN), Naive Bayes (NB) and DT. The best performing classifier was

DT with an accuracy of 93.23% and a false alarm rate of 6.77%. Meftah et al. (2019) [20] used a two-stage classification approach to detect network attacks. The preprocessing involved preparing the UNSW-NB15 dataset, and reducing features using Recursive Feature Elimination (RFE) and RF algorithms. The first stage of proposed solution consist of classifying traffic as normal or intrusion using the best result offered by one of trained models: Logistic regression (LR), GB or Support Vector Machine (SVM). The output is applied to another classification stage, with the aim of improving the performance of the NIDS, using the NB, SVM and DT algorithms. The best accuracy of these classification algorithms was 86.04% for DT. Kasongo and Sun (2020) [16] used the same dataset. The classification algorithms LR, KNN, ANN, DT and SVM were applied, and the XGBoost algorithm was used for the Feature Selection. The binary classification with the KNN algorithm performed the best with an accuracy of 96.76%, while the multiclass classification reached only 82.66% accuracy. Kumar and al. (2021) [18] designed a NIDS that detect normal traffic and classify malicious traffic into 04 categories of network attacks : Denial of Service (DOS), generic, exploit and recognition. The system was trained on the UNSW-NB15 dataset, to which the preprocessing steps have been applied, namely: dataset clustering to reduce the number of attack categories from 9 categories to only 4 the most representative; the gain of information for the feature selection, which made it possible to choose the attributes whose information gain is positive. The researchers started with 22 attributes, calculated the performance of the models (C5, CART, CHAID and QUEST) that were trained on the dataset, and then 13 attributes were selected. The final model recorded an accuracy of 88.92%, but couldn't succeed to detect 5 types of attacks.

Furthermore, some authors used obsolete dataset [5], i.e. KDD-99 . It was the case for Priya and al. (2019) [26] who used LR, RF, GB, and XGBoost algorithms to implement a NIDS trained on the aforementioned dataset. The results obtained in binary classification were quite satisfactory with more than 96.6% in area under the curve.

In [4], a new balanced resampling of outdated KDD-99 [5], namely NSL-KDD, was used as a benchmark dataset to propose a NIDS based on a binary XGBoost-DNN classifier. The feature selection was performed using XGBoost algorithm and reduced dimensionality from 41 to 21 features. The preprocessed dataset was fed to a Deep Neural Network (DNN) classifier. To enhance training, authors used Adam Optimizer and obtained a maximum accuracy of 97%. The model's results were compared on the basis of accuracy to LR, SVM, and NB models and proved its superiority.

Karatas and al. (2020) [15] used the K-Nearest Neighbor (KNN), RF, GB, Adaboost, DT, and Linear Discriminant Analysis (LDA) algorithms to train the recent CSE-CIC-IDS2018 dataset. During the preprocessing phase, the researchers opted for Synthetic Minority Oversampling Technique (SMOTE) to deal with the problem of imbalanced classes. The technique increased the number of minority class instances to the average of class instances. It resulted in an improvement of the average accuracy of the models between 4% and 30%.

Unfortunately, several classes had low accuracy performances, which made it difficult for the proposed classifier to precisely identify the type of attacks.

Alani and Miri (2022) [1] built binary intrusion classification models for IoT using RF, LR, DT and Gaussian Naive-Bayes (GNB) algorithms. In this study, DT was trained on TON_IoT,posemat IoT-23 and IoT-ID datasets. The model recorded an accuracy of 99.62% with a prediction time that was improved by 70%, to less than 0.5$\mu$s, thanks to the reduction of the number of attributes to six using the RFE algorithm.

In [29], a hybrid approach to develop a NIDS was proposed. In their work, Talukder et al. made use of SMOTE for minority classes oversampling, and XGBoost algorithm for feature selection. Furthermore, authors trained ML and DL algorithms, namely RF, DT, KNN, MLP, CNN on the preprocessed KDD-CUP'99 and CIC-MalMem-2022 datasets. Although obtained results were outstanding (above 99% of accuracy), both of used datasets present some shortcomings. In fact, in addition to outdated KDDCUP'99, CIC-MalMem-2022 dataset is designed to develop obfuscated malware detection systems [2], which may not be able to detect other intrusion attacks types, such as network attacks, web attacks,etc.

The table 2 gives performances and limitations of works related to our problematic.

## 4    Proposed Solution

### 4.1    Choice of Dataset

In [25], Pavlov and al. proposed criteria (variety and relevance of attacks, balance of attack types, number of attacks, diversity of protocols, correct labeling of instances, etc.) against which recent datasets have been evaluated. To each criterion, the researchers assigned a coefficient and calculated a score for each dataset. The NF-UQ-NIDS-v2 obtained the highest score, and was chosen for our study. It is a recent dataset, developed by researchers at the University of Queensland Australia in 2021 [27]. These scientists merged the datasets (UNSW-NB15, BoT-IoT, ToN-IoT, CSE-CIC-IDS2018) which were converted into a dataset whose features are those of the Netflow v9, which is widely implemented protocol in network equipment. The dataset contains 75.987.976 instances of which 25.165.295 (representing 33.12%) are normal flows and 50.822.681 (representing 66.88%) are malicious flows with 20 classes of attacks (MITM, Ransomware, DDOS, Shellcode, Backdoor,etc.).

## 4.2   Dataset Observations and Preprocessing

The dataset underwent careful examination to find any anomalies that might hinder the training process:

– No missing values were found;
– The dataset includes features that are host-specific such as the src-ip and dst-ip, which have been removed [21];
– All remaining data fields have numerical values, except Attack feature;
– Three attack types have been removed from the dataset:
  • Worm attacks due to insufficient training records;
  • Password attacks class because it contains two separate modus operandi: brute force technique and sniffing;
  • Exploit attacks, since they are responsible for class overlap phenomenon [32]. This finding is further corroborated by the fact that exploits include separate techniques like brute force, XSS, DOS....

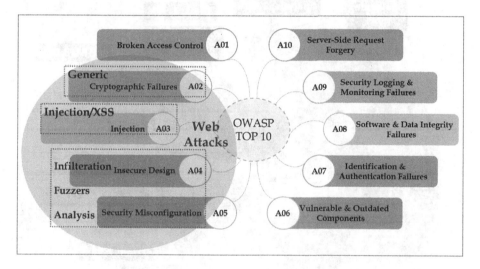

**Fig. 3.** Mapping between web attacks category classes and OWASP Top 10 attacks

Furthermore, a grouping of web attacks (Injection, XSS, Infiltration, Fuzzers, Generic, and Analysis) into one category, Fig. 3, was carried out using definitions and the matches between these classes and the web attacks highlighted by Open Web Application Security Project (OWASP) Top 10: 2021 [24].

### 4.3   Dataset Balancing

According to the above-mentioned comparative study [25], the attack type balance ratio is 0,477. To deal with this shortcoming, we operated a transformation of the dataset using K-means clustering-based sampling method. In fact, for every type of attack, the records were grouped into clusters. Moreover, a new attribute 'cluster label' was added temporarily to the dataset to affiliate each instance to its cluster. After running the sampling function, which take into consideration the number of output records, a random sampling was applied to each cluster. Indeed, this technique is well known for preserving the quality of sampling, since all eliminated instances are generally redundant ones within the clusters. As a result, the sampled dataset become highly balanced, with a ratio of 1:1 between the 'attack' class and the 'benign' class and fair proportions between the attack types. Figure 4 depicts the proportion of classes in the sampled dataset.

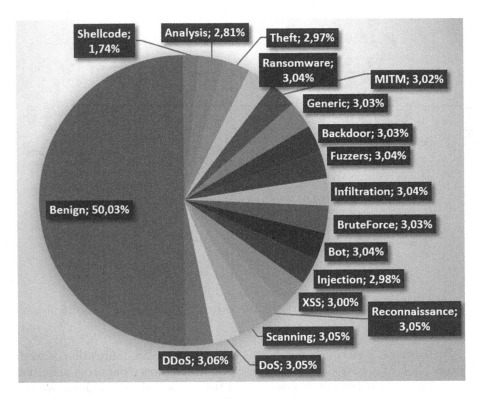

**Fig. 4.** Proportion of classes after applying K-means clustering-based sampling

## 4.4 Work Tools and Environment

Proposed models were implemented using Python programming language, and Scikit-learn, Numpy and pandas libraries. All our experiments were executed on a workstation that has the following specifications: Intel(R) Core(TM) i7-6600U CPU @ 2.60GHz (4 CPUs), 16Go in RAM memory.

# 5 Results and Discussions

## 5.1 Evaluation Metrics

Intrusion detection is a classification problem, wherein classifiers are evaluated by mapping actual to predicted values using confusion matrix, table 3. True Positive (TP) is an actual attack sample correctly predicted , False Negative (FN) is an attack sample predicted as normal, True Negative (TN) is a normal sample correctly predicted, False Positive (FP) is a normal sample predicted as attack.

The following evaluation metrics, based on confusion matrix, were adopted for our study:

**Table 3.** Confusion matrix for intrusion detection

|  |  | Predicted class | |
|---|---|---|---|
|  |  | *Normal* | *Attack* |
| **Actual Class** | *Normal* | TN | FP |
|  | *Attack* | FN | TP |

- Accuracy: the number of samples correctly predicted divided by the total number of predictions:

$$Accuracy = \frac{TP + TN}{TP + TN + FP + FN} \tag{1}$$

- Precision: the proportion of actual attacks detected among all triggered alerts ( actual attacks and normal traffic detected as an attack):

$$Precision = \frac{TP}{TP + FP} \tag{2}$$

- Recall: the proportion of actual attacks detected among all attacks (the ones detected and the ones not detected ):

$$Recall = \frac{TP}{TP + FN} \tag{3}$$

- F1 score: a harmonic mean of the precision and recall. Hence, a high F1-score means that the two metrics are high:

$$F1\_score = 2 \cdot \frac{Precision \cdot Recall}{Precision + Recall} \tag{4}$$

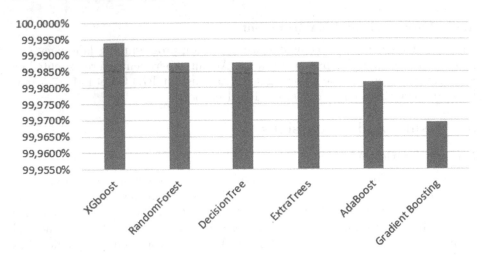

**Fig. 5.** F1 score for binary classification

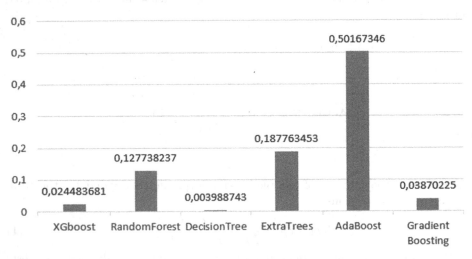

**Fig. 6.** Prediction time (in seconds)for binary classification

## 5.2   Binary Classification

The purpose of binary classification is to detect whether an intrusion took place (
1: Attack) or not (0 : Normal). For our study, binary classification yielded excellent performances. All the six algorithms recorded an accuracy and F1 score above 99.96% with prediction time under 0.5 s. No FN was registered in the majority of the proposed algorithms and FPs was under 2 instances for all algorithms except Adaboost.

A summary of results for binary classification is given in table 4. As can be seen in the Fig. 5, XGBoost algorithm was the most successful algorithm with an F1 score of 99.99%. Decision Tree algorithm present the advantage of having a prediction time that is less than 4ms, Fig. 6, while preserving an accuracy and F1 score of 99.98%.

**Table 4.** Summary of results for binary classification

| Classifier | F1 score | FP | FN | Training time | Prediction time |
|---|---|---|---|---|---|
| XGBoost | 99,9939% | 1 | 0 | 4,1803 | 0,0244 |
| Decision Tree | 99,9878% | 2 | 0 | 1,1250 | 0,0039 |
| Random Forest | 99,9878% | 2 | 0 | 5,4591 | 0,1277 |
| ExtraTrees | 99,9877% | 1 | 1 | 5,8465 | 0,1877 |
| AdaBoost | 99,9816% | 1 | 2 | 9,7301 | 0,5017 |
| Gradient Boosting | 99,9695% | 5 | 0 | 35,7075 | 0,0387 |

## 5.3  Multiclass Classification

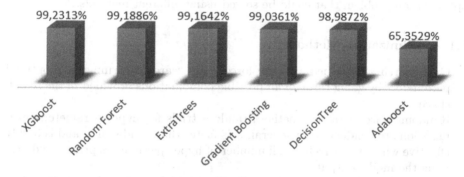

**Fig. 7.** Accuracy of multiclass classification

With the exception of Adaboost, which performed below 66% in accuracy, the metrics of the other algorithms performed well, as can be seen in Fig. 7. It can be concluded, after the analysis of the results, that the most successful model is XGboost which recorded the best performance (99.23% in accuracy, 0.13 s in prediction time). Among the algorithms with the best performance, i.e. accuracy higher than 99%, XGBoost stands out for its prediction time which is the shortest, Fig. 8.

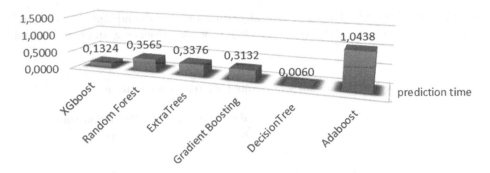

**Fig. 8.** Prediction time (in seconds) for multiclass classification

# 6    Hyper-parameters Optimization

Most of machine learning algorithms need hyper-parameters to be specified before learning phase [12]. Examples of hyper-parameters include number of hidden layers in neural networks , number of decision trees (n_estimators), and maximum depth of the decision tree (Max_depth). Analytically or numerically finding optimal combination of hyper-parameters for better performances is an optimization problem that could be solved using different methods.

## 6.1    Optimization Methods

- Grid Search : Described as brute force, Grid search determines the optimal parameters by going through all possible combinations provided in the search space.
- Random Search : In this method, random trials for hyper-parameters optimization are conducted. In general, it is faster than Grid search and is highly effective when a relatively small number of hyper-parameters primarily determine the model output.
- Genetic algorithm optimization : Similar to biological evolution, the algorithm start with a randomly generated population of individuals (hyper-parameters in search space). Next, a selection process and recombination of hyper-parameters is performed, then evaluated to choose the ones who 'survived' and could adapt to changes. The process is repeated till it reach final destination criterion [19].
- Bayesian optimization : Bayesian optimization is an iterative algorithm based on two elements: the surrogate model and the acquisition function [31]. A surrogate model is an approximation of a computationally expensive function. The evaluation of the surrogate model is very quick, which allows to estimate the value of the cost function without having to actually evaluate it. After obtaining the predictive distribution of the surrogate model, the acquisition function choose to use different parameters by resorting to trade-offs between exploration and exploitation. Exploration involves sampling instances in areas

that have not been sampled, while exploitation involves sampling in currently promising regions where the global optimum is most likely to occur, based on the posterior distribution. Tree Parzen Estimator (TPE) is a surrogate model that converts parameters intervals into a nonparametric density distribution. The parameter range can be represented by a uniform distribution, a discrete uniform distribution (q-uniform) or a logarithmic uniform distribution (log-uniform). Therefore, TPE is more flexible than traditional Bayesian optimization, and was chosen in our study .

## 6.2   Optimization Results

Since XGBoost is the algorithm that provided the best results, its hyper-parameters, namely n_estimators, max_depth, and learning rate, were optimized. We chose Bayesian Optimization Tree Parzen Estimator BO-TPE, as it is suitable for XGBoost and has proven its effectiveness in two studies [6,30].

We implemented a python function, with the help of hyperopt library, which returned the best hyper parameters, table 5. After applying the best hyper-parameters, the performance of XGBoost in multiclass classification achieved an accuracy of 99.28% and a precision of 99.29%.

**Table 5.** BO-TPE hyper-parameter optimization parameters and optimal values

| Hyper-parameter | Space | step | Distribution | Optimal value |
|---|---|---|---|---|
| n_estimator | [50,200] | 10 | Quniform | 180 |
| max_depth | [2,15] | 1 | Quniform | 5 |
| learning_rate | [0.05,0.5] | | normal | 0.1056 |

# 7   Conclusion and Future Work

The results obtained are promising in comparison with the most recent scientific articles dealing with NIDS conception. Indeed, the binary classification recorded an accuracy and an F1 score of 99.99% using XGBoost, and a prediction time that is less than 4ms for Decision Tree while preserving an accuracy of 99.98%. In addition, the multiclass classification, which makes it possible to categorize the intrusion into 13 types of attacks, recorded an accuracy of 99.28% and a precision of 99.29% after using the BO-TPE optimization method, with a prediction time of 240ms. The quality of the results obtained is justified by the choices made during the preprocessing phase. First, the transformation of the NF-UQ-NIDS-v2 dataset into a highly balanced dataset, using the sampling method based on K-means clustering before training the models. Secondly, through the exploration of classes, the problem of class-overlap has been mitigated by eliminating certain classes that suffer from this phenomenon. Finally, a grouping of Web attacks classes was performed using the correspondences between the dataset classes and Web attacks highlighted by OWASP.

We believe our DT model could be implemented in resources-constrained IoT devices and Cyber Physical Systems (CPS), such as Unmanned Aerial Vehicle UAV-mounted Base Station, to detect intrusion as quickly as possible. Additionally, the XGBoost model would be a perfect choice for categorizing cyberattacks.

In future works, improving the multiclass model's performances by using other optimization methods would be examined. In addition, the focus will also be toward exploring Deep Learning algorithms and seeking to minimize attributes, through feature selection to improve prediction and training time.

# References

1. Alani, M.M., Miri, A.: Towards an explainable universal feature set for IoT intrusion detection. Sensors **22**(15), 5690 (2022). https://doi.org/10.3390/s22155690, https://www.mdpi.com/1424-8220/22/15/5690
2. Carrier, T., Victor, P., Tekeoglu, A., Lashkari, A.H.: Malware Memory Analysis | Datasets | Canadian Institute for Cybersecurity | UNB — unb.ca. https://www.unb.ca/cic/datasets/malmem-2022.html. Accessed 12 Oct 2023
3. Cichonski, P., Millar, T., Grance, T., Scarfone, K.: Computer security incident handling guide(800-61-revision 2). Nat. Inst. Stand. Technol. **10** (2012)
4. Devan, P., Khare, N.: An efficient XGBoost-DNN-based classification model for network intrusion detection system. Neural Comput. Appl. **32**, 12499–12514 (2020)
5. Divekar, A., Parekh, M., Savla, V., Mishra, R., Shirole, M.: Benchmarking datasets for anomaly-based network intrusion detection: KDD CUP 99 alternatives. In: 2018 IEEE 3rd International Conference on Computing, Communication and Security (ICCCS), pp. 1–8. IEEE (2018)
6. Dong, H., He, D., Wang, F.: SMOTE-XGBoost using tree parzen estimator optimization for copper flotation method classification. Powder Technol. **375**, 174–181 (2020). https://doi.org/10.1016/j.powtec.2020.07.065, https://linkinghub.elsevier.com/retrieve/pii/S0032591020306896
7. Dunsin, D., Ghanem, M.C., Ouazzane, K., Vassilev, V.: A comprehensive analysis of the role of artificial intelligence and machine learning in modern digital forensics and incident response. arXiv preprint arXiv:2309.07064 (2023)
8. El Mrabet, Z., El Ghazi, H., Kaabouch, N.: A performance comparison of data mining algorithms based intrusion detection system for smart grid. In: 2019 IEEE International Conference on Electro Information Technology (EIT), pp. 298–303. IEEE (2019)
9. El Mrabet, Z., Ezzari, M., Elghazi, H., El Majd, B.A.: Deep learning-based intrusion detection system for advanced metering infrastructure. In: Proceedings of the 2nd International Conference on Networking, Information Systems & Security. NISS19, Association for Computing Machinery, New York, NY, USA (2019). https://doi.org/10.1145/3320326.3320391
10. El Mrabet, Z., Kaabouch, N., El Ghazi, H., El Ghazi, H.: Cyber-security in smart grid: survey and challenges. Comput. Electr. Eng. **67**, 469–482 (2018)
11. Fausto, A., Gaggero, G.B., Patrone, F., Girdinio, P., Marchese, M.: Toward the integration of cyber and physical security monitoring systems for critical infrastructures. Sensors **21**(21) (2021). https://doi.org/10.3390/s21216970

12. Florea, A.C., Andonie, R.: Weighted random search for hyperparameter optimization. arXiv preprint arXiv:2004.01628 (2020)
13. González-Granadillo, G., González-Zarzosa, S., Diaz, R.: Security information and event management (SIEM): analysis, trends, and usage in critical infrastructures. Sensors **21**(14), 4759 (2021)
14. Henriques, J., Caldeira, F., Cruz, T., Simões, P.: Combining k-means and XGBoost models for anomaly detection using log datasets. Electronics **9**(7) (2020). https://doi.org/10.3390/electronics9071164, https://www.mdpi.com/2079-9292/9/7/1164
15. Karatas, G., Demir, O., Sahingoz, O.K.: Increasing the performance of machine learning-based IDSS on an imbalanced and up-to-date dataset. IEEE Access **8**, 32150–32162 (2020)
16. Kasongo, S.M., Sun, Y.: Performance analysis of intrusion detection systems using a feature selection method on the UNSW-NB15 dataset. J. Big Data **7**(1), 105 (2020). https://doi.org/10.1186/s40537-020-00379-6
17. Koroniotis, N., Moustafa, N., Sitnikova, E., Slay, J.: Towards developing network forensic mechanism for botnet activities in the IoT based on machine learning techniques. In: Hu, J., Khalil, I., Tari, Z., Wen, S. (eds.) MONAMI 2017. LNICST, vol. 235, pp. 30–44. Springer, Cham (2018). https://doi.org/10.1007/978-3-319-90775-8_3
18. Kumar, V., Das, A.K., Sinha, D.: UIDS: a unified intrusion detection system for IoT environment. Evolution. Intell. **14**(1), 47–59 (2019). https://doi.org/10.1007/s12065-019-00291-w
19. Liashchynskyi, P., Liashchynskyi, P.: Grid search, random search, genetic algorithm: a big comparison for NAS. arXiv preprint arXiv:1912.06059 (2019)
20. Meftah, S., Rachidi, T., Assem, N.: Network based intrusion detection using the UNSW-NB15 dataset. Int. J. Comput. Digit. Syst. **8**(5), 478–487 (2019)
21. Moustafa, N.: A new distributed architecture for evaluating AI-based security systems at the edge: Network TON_IoT datasets. Sustain. Cities Soc. **72**, 102994 (2021). https://doi.org/10.1016/j.scs.2021.102994, https://linkinghub.elsevier.com/retrieve/pii/S2210670721002808
22. Nilă, C., Patriciu, V.: Taking advantage of unsupervised learning in incident response. In: 2020 12th International Conference on Electronics, Computers and Artificial Intelligence (ECAI), pp. 1–6. IEEE (2020)
23. Nilă, C., Apostol, I., Patriciu, V.: Machine learning approach to quick incident response. In: 2020 13th International Conference on Communications (COMM), pp. 291–296 (2020). https://doi.org/10.1109/COMM48946.2020.9141989
24. OWASP Top 10:2021 — owasp.org. https://owasp.org/Top10/. Accessed 07 Oct 2023
25. Pavlov, A., Voloshina, N.: Dataset selection for attacker group identification methods. In: 2021 30th Conference of Open Innovations Association FRUCT, pp. 171–176 (2021). https://doi.org/10.23919/FRUCT53335.2021.9599966
26. Priya, S., Sahu, B.K., Kumar, B., Yadav, M.: Network intrusion detection system using XGBoost. Int. J. Eng. Adv. Technol. **9**(1), 4070–4073 (2019). https://doi.org/10.35940/ijeat.A1307.109119, https://www.ijeat.org/portfolio-item/A1307109119/
27. Sarhan, M., Layeghy, S., Portmann, M.: Towards a standard feature set for network intrusion detection system datasets. Mob. Netw. Appl. **27**(1), 357–370 (2022). https://doi.org/10.1007/s11036-021-01843-0
28. Scarfone, K.A., Mell, P.M.: Guide to intrusion detection and prevention systems (IDPS). Technical report NIST SP 800-94, National Institute of Standards

and Technology, Gaithersburg, MD (2007). https://doi.org/10.6028/NIST.SP. 800-94, https://nvlpubs.nist.gov/nistpubs/Legacy/SP/nistspecialpublication800-94.pdf, edition: 0

29. Talukder, M.A., et al.: A dependable hybrid machine learning model for network intrusion detection. J. Inf. Secur. Appl. **72**, 103405 (2023). https://doi.org/10.1016/j.jisa.2022.103405, https://www.sciencedirect.com/science/article/pii/S2214212622002496

30. Yang, L., Moubayed, A., Shami, A.: MTH-IDS: a multi-tiered hybrid intrusion detection system for internet of vehicles. IEEE Internet Things J. 616–632 (2022). https://doi.org/10.1109/JIOT.2021.3084796, http://arxiv.org/abs/2105.13289, arXiv:2105.13289 [cs]

31. Yang, L., Shami, A.: On hyperparameter optimization of machine learning algorithms: theory and practice. Neurocomputing **415**, 295–316 (2020). https://doi.org/10.1016/j.neucom.2020.07.061, https://www.sciencedirect.com/science/article/pii/S0925231220311693

32. Zoghi, Z., Serpen, G.: UNSW-NB15 computer security dataset: analysis through visualization. arXiv preprint arXiv:2101.05067 (2021)

# Author Index

Printed in the United States
by Baker & Taylor Publisher Services